Farming systems development and soil conservation

David Norman
and
Malcolm Douglas

2007
Published by Arrangement with the
Food and Agriculture Organization
of the United Nations
by
DAYA PUBLISHING HOUSE
Delhi - 110 035
for Sale and Distribution only in South-East Asia

"This book was originally published by the Food and Agriculture Organization of the United Nations (FAO) as FAO Farm Systems Management Series No. 7."

First Indian Edition 2007

ISBN 81-7035-512-5

Published by : **Daya Publishing House**
1123/74, Deva Ram Park
Tri Nagar, Delhi - 110 035
Phone: 27383999
Fax: (011) 23244987
e-mail : dayabooks@vsnl.com
website : www.dayabooks.com

Showroom : 4760-61/23, Ansari Road, Darya Ganj,
New Delhi - 110 002
Phone: 23245578, 23244987

Printed at : **Salasar Imaging Systems**
Delhi - 110 35

PRINTED IN INDIA

PREFACE

This publication has been jointly promoted and produced by the Farm Management and Production Economics Service, Agricultural Services Division, and the Soil Resources, Management and Conservation Service, Land and Water Development Division, Food and Agriculture Organization (FAO), Rome, Italy. It is based on the contribution of two main authors, namely:

- **D.W. Norman, Professor Agricultural Economics, Kansas State University, USA. (KSU contribution number 94-234-B)**

- **Malcolm Douglas, Consultant/Trainer Land Husbandry, York, UK.**

They had prepared two earlier papers, which provided the foundation of the present publication. These were:

- **A Framework for the Development of Conservation Effective Farming Systems, 1992 (M.G. Douglas).**

- **Soil Conservation: Using Farming Systems Development as an Aid, 1991 (D.W. Norman).**

In addition, this publication draws freely on the work of others including the **Framework for Land Evaluation** and **Farming Systems Development** approaches of FAO, the **Diagnosis and Design** approach of ICRAF, the publications of IIED and others on **Rapid Rural Appraisal** and **Participatory Rural Appraisal**, and an FAO Discussion Paper (March 1990) by A.C. Vlaanderen titled **Changes in Soil Conservation: Implications and Applications**.

A number of people have produced constructive comments on the papers produced earlier and also on this publication. In particular, the suggestions and advice of D. Carandang, T.R. Frankenberger, B. Felizardo, K. Friedrich, R. Gallacher, K.C. Lai, R. Ranola, D. Sanders, T.F. Shaxson, K. Shepherd, J.R. Williams, and F.W. Worman, are gratefully acknowledged.

The objectives of this publication are to:

- Provide a reasoned discussion on why it is desirable to incorporate elements of the farming systems development (FSD) approach in handling issues relating to the new approach to soil conservation.

- Develop a conceptual framework for the incorporation of the FSD approach.

- Outline some of the FSD techniques that could be useful in identifying, designing and testing/implementing relevant soil conservation strategies.

This publication addresses as readership subject-matter specialists in soil conservation, who are interested in their work becoming more effective by integrating elements of the FSD approach into it.

TABLE OF CONTENTS

LIST OF TABLES

LIST OF FIGURES

1. INTRODUCTION

1.1 THE PROBLEM OF SOIL EROSION

It has recently been estimated that millions of hectares of cultivated land are lost to agricultural production each year because of soil degradation [ICRISAT, 1986; FAO, 1992]. This is happening because the rapidly growing world population is demanding ever increasing quantities of food, fibre, and fuel from the land. As a result, land is being farmed more intensively. In addition, increasing quantities of new land being brought into cultivation are of poorer quality and less productive than land previously being used in production, therefore, exacerbating the problems of soil degradation. Not surprisingly, sustainability of the land resource base has become of great concern in recent years.

This concern can been highlighted in another way, that is, in terms of regeneration. Although very dependent on the type of soil management and the level of organic matter, estimates indicate that almost 25 percent of the cropland in the USA will lose 2.54 centimetres of soil every 30 years [Napier and Forster, 1982]. Under ideal conditions, it takes about 30 years to produce 2.54 centimetres of topsoil, but usually it takes much longer [Pimentel et al, 1976]. Because restoring topsoil is very difficult, trying to prevent erosion, that is, ensuring sustainability, is much more appropriate than trying to reclaim the land once the topsoil has been lost.

Few quantitative estimates are available of the rates of soil loss in the low income countries. Nevertheless, these rates are likely to be much greater in many such countries than in the USA [Anderson and Thampapillai, 1990]. This is because, in many low income countries, no policies exist to encourage soil conservation, while rapidly increasing populations are putting pressure on the land resource base. Environmentally sound traditional practices attuned to low population densities have been unable to adjust rapidly enough to the decreasing land per resident ratios, resulting in practices that are environmentally damaging [FAO, 1986]. One definitive study was undertaken in Zimbabwe. It involved estimation of soil nutrient losses from experimental data collected in the late 1950s and early 1960s [Stocking, 1986]. The results indicated that there was little difference in predicted soil nutrient losses (i.e., soil, phosphorous, and organic carbon) with variations in season, soil type (except for one type), crop, or degree of erosion. However, a major concern was the finding that the enrichment ratios[1] on average were about 2.5, being highest where runoff was greatest. Stocking estimated that these nitrogen and phosphorous losses from arable land were equivalent to three times the level of total fertilizer application in Zimbabwe in 1984/85. When expressed in financial terms, the estimated cost from erosion, depending on its level, varied from $US20-50/ha on arable lands and $US10-80/ha on grazing lands.[2] Thus, erosion appears to have a high hidden cost for the Zimbabwe economy and, therefore, is undermining its resource base for the future. Such on-site costs are likely to be even higher, if other results of erosion are taken into account.

However, the costs associated with soil erosion are more complicated than the above discussion suggests. Such costs can be classified into two broad groups [Napier, 1987]:

---

1 This measures how much more concentrated a nutrient is in the eroded soil than it is in the soil from which it is taken. Such ratios do not include losses of nutrients dissolved in runoff water.

2 These estimates were based on the equivalent cost of nitrogen and phosphorus fertilizers required to replace the lost nutrients. 1985 rates of foreign exchange were used.

- ***On farm site costs.*** These are problems associated with loss of soil nutrients and productivity. The costs involved are usually borne by the farmers themselves. Such costs provide personal incentives for adoption of practices to combat soil erosion.

- ***Off farm site costs.*** These consist of problems that are exported off the farmers' land, such as general gully erosion, sedimentation of waterways and reservoirs, and degradation and contamination of drinking water. The costs of such problems are usually shifted to other members of society. Consequently, few personal incentives exist to implement strategies for combating such costs, unless reducing on-site costs simultaneously reduces off-site costs. However, such costs are of major concern to society. Because the pricing system does not automatically reflect such external costs, governments in high income countries are increasingly trying to apportion responsibility for meeting such costs through regulations that shift some, if not all, of the cost burden back to those initially responsible for creating the problem.

Although, in high income countries, considerable attention is now being focused on off-site damages, little attention, particularly research, has been devoted to looking at such costs in low income societies [Napier, 1989]. Napier has suggested that off-site costs of soil erosion are likely to be more important than on-site damages in low income countries. He attributes this to the fact that the degree of soil erosion tends to be higher and some chemicals are used that can contaminate water and are banned in high income countries. However other authors suggest that the on-site costs of soil erosion may greatly exceed off-site ones.[3]

---

3 Analysis of soil erosion on Java [Magrath and Arens 1989] estimated annual off-site costs at US$25.6-91.2 million, as compared to $315 million for on-site costs (i.e., productivity losses).

Finally, low income countries, faced with very limited resources and rapidly increasing populations, emphasize increasing current agricultural productivity. Consequently, little consideration can be given to erosion that has already developed or that will develop in the future. Many of the technologies that have been introduced to date tend to exacerbate the trend towards erosion and unsustainability by dominating environments through the use of machinery, chemicals, irrigation, and other capital-intensive inputs [Conway, 1987; Francis and Hildebrand, 1989]. To preserve sustainability and hence prevent erosion, technologies must be compatible with the environments.

So, what are the impacts of erosion? Stocking [1986] has divided the impacts into two parts, as follows:[4]

- The effects on how specific soils can be used include:

    - Irregular removal of soil from a field causes variations in the return from the uniform application of fertilizer and herbicide. It also can affect the use of machinery.

    - Timing of farming operations can be severely affected because crusted and compacted soils make land difficult to prepare, particularly with draught animals. Late planting and delayed germination often occur, adversely affecting yields.

    - Off-site damage, such as sedimentation of drainage channels and reservoirs, is common. Results can include reduction in water storage capacities, flooding, and inundation of fertile top soils.

    - Changes in land use and the farming systems are induced, usually to

---

4 Greater detail on specific impacts can be found in FAO [1984].

accommodate the system to a lower level of productivity. Possible changes include growing crops that can tolerate less favourable growing conditions, which are typically less palatable and nutritious, moving to more of a pastoralist system, and changing from large to small stock.

- The effects on soil characteristics and productive potential are as follows:

  - Nutrients (i.e., particularly nitrogen and phosphorus but also other nutrients associated with cation exchange or with organic matter) are lost in the eroded material in amounts proportionately higher than those in the original soil.

  - Toxicities and pH-related deficiencies may be exacerbated. Aluminum toxicity can be a problem in strongly acid soils. The higher aluminium levels found in the subsoil of some tropical soils may be a factor in explaining yield decline following loss of topsoil by erosion. Aluminium toxicity and acidification in turn, make other nutrients unavailable to plants.

  - Water availability to plants decreases as a result of one or more of the following: reduced soil depth; loss of water-holding capacity through erosion of clays and organic matter and/or reduced rooting volume because of the unearthing of a stone, sodic, or lateritic layer; or some other hindrance to rooting.

  - The structural stability of the soil collapses through an increase in bulk density, raindrop impact, loss in organic matter, etc. Consequently, puddling, sealing and compaction of the soil occurs.

Therefore, erosion results from mismanagement of arable land, rangeland, and forested areas. This usually consists of removing too much; returning too little; and/or cultivating, grazing, or cutting too frequently [FAO, 1986]. The overall impact of erosion is a decrease in the size of the A horizon, reducing the supply of water and nutrients available for plant growth. This implies that maintenance of yield levels requires ever-increasing quantities of external inputs (e.g., fertilizer) which may be uneconomical. Therefore, it is increasingly obvious that a major crisis is developing. Soil erosion has already devastated, sometimes irreversibly, large areas of many low income countries. In addition, in other areas, yields are slowly declining or are only being maintained, usually temporarily, through massive quantities of external inputs. Both the expense and lack of success in past public, and sometimes private, sector initiatives to implement soil erosion control initiatives has led to increasing interest in the issue of sustainability of the natural resource base. This reorientation means including design and implementation of strategies that prevent the problem of soil erosion developing in the first place. In other words strategies need to be implemented that encourage good land husbandry.

1.2 CHANGES IN THINKING ABOUT SOIL CONSERVATION OR LAND HUSBANDRY

As implied above, parallel to the growing interest in sustainability issues has been increased self-examination by those responsible for research on, and implementation of, soil conservation projects and programmes. A large number of reasons have been given why soil conservation projects have not been successful. Some of those that have been suggested include the following [Sanders, 1988; Shaxson et al, 1989]:

- Prevention of soil erosion has been considered an end in itself.

- Soil scientists and engineers have perceived the main effect of soil erosion to be soil loss, rather than considering its effects on production.

- The farmer often has been viewed as one of the factors to be changed or, in more extreme cases, as constituting the problem to be overcome.

- Farmers are seldom interested in programmes that have few short-term benefits (i.e., that offer few personal incentives), uncertain long-term benefits and involve additional work, costs, and risks.[5]

- Often, the problems farmers have in cooperating in soil conservation programmes have not been sufficiently appreciated. Reasons can be very diverse and often have their roots in socioeconomic circumstances.

In essence, perhaps part of the problem has often been the belief that the basic assumptions[6] on which many soil conservation projects have been based are appropriate or right and that farmers who don't willingly participate in their implementation are wrong. In fact the farmers may be right and the technical solutions proposed may have severe limitations or be completely inappropriate. Therefore, lack of past success has stimulated many individuals working in the soil conservation area to develop a new set of principles on which soil conservation should be based. Consequently, there is now general agreement on the following [Hudson, 1988; Shaxson et al, 1989; Douglas, 1990; Douglas 1991A]:

- Loss of soil productivity is much more important than the loss of soil itself. Therefore, soil conservation must be an integral part of general

---

5 Security of land tenure will also influence the farmer's willingness to incur short term costs (i.e., including foregone benefits) if the expected benefits are, at best, long-term.

6 Typical assumptions used have been that: farmers perceive soil erosion as a problem and will invest effort in tackling it; that soil degradation can be combatted by a purely technocratic approach; and that farmers can adopt the technologies proposed and that these will prevent further degradation.

agricultural development and should start with improved farming and improved production -- and not be considered as a separate activity.

- Erosion is a consequence of how the land is used and is itself not the cause of soil degradation. Land degradation, therefore, should be prevented before it arises, instead of attempting to cure it afterwards -- that is, sustaining the productivity of the soil resource is critically important.

- Land has been studied too much in soil conservation programs, whereas the land user (i.e., farmer) has been studied too little. Therefore, soil conservation activities must be 'bottom-up' rather than 'top-down' orientation and planned and implemented from the outset, with the knowledge, cooperation and involvement of the farmers and lo communities. At the same time, to be attractive to farmers, an proposed soil conservation activity must provide farmers short-term benefits (i.e. incentives).

- Plant yields are reduced more by a shortage or excess of soil moisture than they are by loss of soil. Hence there should be more emphasis on rain water management, particularly water conservation, and less on soil conservation *per se*. The primary focus should be on reducing run-off (i.e., by encouraging infiltration) before trying to control its overland flow. Consequently, agronomic measures (i.e., tillage, crop management practices) are potentially more significant than mechanical measures in preventing erosion and run-off.

- Help in soil conservation is more successful when applied through long-term programmes rather than through short, fixed-term projects.

The above points indicate fundamental changes in both the philosophy and the approach used by soil conservationists. Philosophically a major part of the new approach is perhaps more aptly called land husbandry[7] rather than soil conservation. The change in approach involves moving from looking at erosion in terms of ***what is happening*** to ***why it is happening*** [Shaxson et al, 1989]. The former approach implies that only the symptoms of erosion are examined. However, examining the 'why' component is potentially much more useful in extracting the causes underlying the erosion (e.g., what are the biophysical and socioeconomic factors that have encouraged, or contributed to the present degradation). An understanding of these provides the basis for designing appropriate strategies to correct them and/or to prevent them from arising in the future.

By now it should be obvious that the key to effective soil conservation is good land management (i.e., land husbandry. The way in which land is managed is the result of a complex interaction between the needs, constraints, and potentials of farmers within the environment (i.e., biophysical and socioeconomic) in which they operate. Therefore, the farmer and his/her environment[8] should be the focal points of every soil conservation programme.

---

7 The concept of husbandry is widely understood when applied to crops and animals. As a concept signifying understanding, management and improvement, it is equally applicable to land. Soil conservation has a somewhat negative image for farmers with its implied need to control and restrict present land use so as to preserve the soil for the future. The concept of land husbandry should prove more acceptable to farmers with its more positive image of managing and improving the present use of the land for productive purposes on a sustainable basis (i.e., husbanding the soil resource) [Douglas 1992].

8 The strictly correct term would be farming household or farming family. In this publication, these terms plus farmer and land user are used interchangeably and mean the same thing.

The above ideas have been articulated for some time but have not yet been applied practically on a large scale. Perhaps the main reason for the delay is that these new ideas imply major changes for soil conservation (i.e., now land husbandry) specialists, changes that are not easy to put into practice. Putting these ideas into practice will require new knowledge and skills and will also involve developing some new methods and techniques.

1.3 IMPLICATIONS CONCERNING THE NEW APPROACH TO SOIL CONSERVATION

There are a number of implications for implementing the new approach to soil conservation. Some of the major ones are as follows:

- Good land husbandry should be an integral part of agricultural development[9] and be the result of improved farming. Therefore, it is essential to understand the actual farming system in order to effectively design and fit in relevant conservation measures. Developing conservation effective measures under such circumstances requires a systems perspective and the need for increased flexibility in thinking.

- Whenever possible, soil erosion should be prevented rather than just cured. However, to design appropriate strategies for prevention or curing, it is important to know the causes of the soil erosion problems and address these, rather than simply treating the symptoms.

- Conservation programmes should be planned with the full knowledge and cooperation of the farmer. Consequently, specialists must be able to communicate effectively with farmers and to obtain ideas of their views,

---

[9] That is well integrated into farm policies and programmes.

perceptions, and priorities. These should be taken into account in developing any strategies for improvement.

- Farmers need to be convinced that there are short-term benefits to be obtained from change (i.e., have an incentive to alter present practices). Thus, it is necessary to address the farmers' immediate needs, which are likely to revolve around increasing current production and/or increased stability/security of subsistence. This requires production strategies that are not only conservation effective but which will also pay-off in the short run. However, if this is not possible, it is important to try and link measures that will provide short term production/welfare benefits for farmers with those required for conservation purposes, which are likely to have a pay-off only in the long-run.

- Long-term programmes are more effective than short-term programmes. Therefore, it follows that the process of change will be slow and long-term in nature and will involve constant adjustments and several iterations.

Soil conservation experts are familiar with classifying soils, analyzing climatic data, and constructing physical structures to protect the land against further erosion. However, the above implications indicate that additional skills are required. For example, the new approach to soil conservation will require being able to:

- Effectively communicate and interact with farmers, through work related activities.

- Classify and analyze farming systems.

- Determine the causes of soil degradation, understand the problems that result, and devise strategies (i.e., particularly, whenever possible, promoting short-term benefits) that will prevent it from arising.

- Collect, analyze, and evaluate socioeconomic and agronomic data, through studies, surveys, and trials.

Having concluded that effective soil conservation or land husbandry will require a more holistic approach than has been used to date, two major issues need resolving.

- Is there a methodology that can be used/adapted to provide a framework that enables this holistic approach to be adopted?
- How can the inputs of the different disciplines be incorporated practically to make the new approach to soil conservation work effective?

The basic premise underlying this publication is that we believe that the rapidly evolving work in ecological sustainability and the methodologies developed for farming systems research (FSR) or farming systems development (FSD)[10] provide good starting points for addressing these issues.

1.4 OBJECTIVES AND CONTENTS OF THE PUBLICATION

The objectives of this publication are to:

- Provide a reasoned discussion on why it is desirable to incorporate

---

[10] Other commonly used names for this type of approach are farming systems research and extension (FSR/E), on-farm research with a farming systems perspective (OFR/FSP), farming systems approach to research (FSAR), agroforestry diagnosis and design (D&D), etc.

elements of the FSD approach in handling issues relating to the new approach to soil conservation.

- Develop a conceptual framework for the incorporation of the FSD approach.

- Outline some of the FSD techniques that could be useful in identifying, designing and implementing relevant soil conservation strategies.

The purpose of the material in Chapter 2 is to provide a foundation for the more practically oriented information presented in the following chapters. Consequently, Chapter 2 provides some basic conceptual information on ecological sustainability and a brief review of the farming systems approach. The specific approach described is farming systems development (FSD), which was developed by the Farm Management and Production Economics Service, Agricultural Services Division, FAO.

Chapter 3 begins with defining the three major strategies that will need implementation, if soil conservation issues are to be comprehensively addressed. The rest of the chapter deals with some of the major elements of the new soil conservation approach and indicates their compatibility with the FSD approach.

Chapter 4 outlines the conceptual framework involved in implementing the new soil conservation approach through integration with the FSD approach, while Chapter 5 provides some tentative practical guidelines.

The final chapter deals briefly with implementing the new soil conservation approach, which involves incorporation of FSD principles.

This publication is intended to identify the need for soil conservation activities to become more closely identified with those used in applying the

FSD approach. As such, it is more of a 'conceptual' rather than a detailed 'how to do it' document. However, for those interested, there are a number of references given in the publication that provide more detail on the tools and techniques used in the application of the FSD approach. For those more interested in the nuts and bolts of the 'bottom up' farmer participatory approach espoused in the new approach to soil conservation, we suggest glancing very quickly at Chapter 3 followed by a more careful scrutiny of the following chapters.

We have also placed some emphasis in the publication on socioeconomic issues, believing that most of the readers are likely to come from technical and soil conservation backgrounds. Consequently, they are less likely to be familiar with socioeconomic issues, and the tools and techniques that will need to be used in successfully implementing the new approach to soil conservation.

2. SUSTAINABILITY AND THE FARMING SYSTEMS DEVELOPMENT APPROACH

2.1 INTRODUCTION

As indicated in Chapter 1, the more traditional approach to soil conservation was narrowly focused and undertaken in isolation, involved a 'top-down' method, had a predominantly technical orientation, and treated the symptoms of soil erosion primarily through mechanical strategies. In contrast, the more modern soil conservation approach has much more of a systems focus involving an integration of conservation and production strategies. To implement this successfully requires a 'bottom-up' method. This can help in providing an understanding of the causes of erosion, which may have their origin not only in the biophysical but also in the socioeconomic environment. In turn, this understanding provides a rational basis for designing strategies that not only will correct the current causes but also will prevent soil erosion developing in the future. Such a change in emphasis in soil conservation orientation towards ensuring sustainability means that increasing attention is now being placed on biological rather than simply on mechanical strategies.

Therefore, to provide a backdrop for implementing the new approach to soil conservation, the following sections delineate some principles that should help in improving ecological sustainability and a brief overview of FSD that can potentially be very useful in effectively implementing the 'bottom-up' method.

2.2 SUSTAINABILITY

Those concerned about sustainable development now generally agree that this will be achieved only if land use systems are developed that are economically viable in the short run and maintain (i.e., sustain) the natural resource productivity in the long run. However, sustainability is not easy to define operationally, although the term must be intuitively comprehensible because many people use it [Hildebrand, 1989].[11] One definition of sustainability, suggested by Jodha [1990], is the ability of the agricultural system to maintain a certain well-defined level of performance (e.g., output) over time, and if required, to enhance that output without damaging the essential ecological integrity of the system.[12] Because of the involvement of the time factor and the need for the system to respond to changing requirements, sustainability is a dynamic rather than a static concept. In essence, it is concerned with inter-generational equity issues, in contrast to the usual use of the term equitability in reference to intra-generational distribution.

There are a number of different perspectives concerning sustainability. Hart and Sands [1991], in a particularly useful conceptual paper, suggest that one way to integrate these and make them operational is through the concept of sustainable land-use systems (Figure 1). The farm system interacts with the biophysical environment through using resources as inputs in production (e.g., water, nutrients, etc.), and through degrading natural resources as a result of producing farm output. Interaction with the socioeconomic environment occurs through purchase of inputs (e.g., seed, fertilizer, etc.) and consumption and sale of outputs (e.g., grain, off-farm employment of family labour, etc.). Also the decision making process within a farm household concerning the component enterprises of its farm system is strongly influenced by its welfare needs and the cultural environment in

11 Sustainability can be thought of in economic, political, social, cultural, institutional, and ecological terms [Hildebrand, 1989] -- although in this publication the emphasis is on ecological sustainability.

12 Ruttan [1988] has indicated that enhancing is essential.

which it operates. Hart and Sands [1991] suggest that different institutions focus on different characteristics of the system depicted in Figure 1. These include those focusing on:

Figure 1: SUSTAINABLE LAND USE SYSTEM

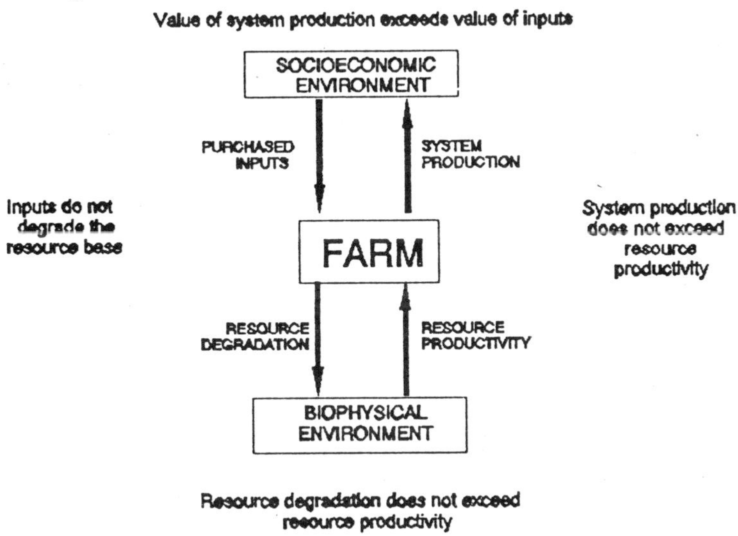

- Achieving short-term economic viability -- the conventional and more traditional approach.

- Using inputs that don't degrade the natural resource base -- the low input (i.e., using fewer purchased inputs) approach.[13]

- Using natural resources while at the same regenerating their productive potential -- the organic farming approach (i.e., whenever possible using renewable resources).

- Maintaining system productivity at a level not exceeding resource productivity -- an applied ecology approach (e.g., forestry, fishing).[14]

Hart and Sands [1991] continue by suggesting that development of sustainable land-use systems will need consideration of all the four system-to-environment relationships. Therefore, they define sustainable land use systems [p. 2] as "those that use both biophysical and socioeconomic resources to produce outputs that the present socioeconomic environment (i.e., today's society) values more than the value of purchased inputs (i.e., the system is economically viable) and that at the same time maintain the productivity of the biophysical environment (i.e., the system is ecologically sustainable)."[15]

---

13 This has to be carefully interpreted since fewer purchased inputs ***per se*** don't necessarily by themselves reduce environmental problems.

14 That is, management of biological resources.

15 In low income countries, many inputs are not purchased. Therefore, use of the opportunity cost approach will be necessary (e.g., valuing family farm labour in terms of what they could have earned in alternative employment).

Hart and Sands [1991] then subdivide the socioeconomic and biophysical environment into local and regional systems (Figure 2). The farm systems are farm households (i.e., which constitute part of the local socioeconomic system) and the crop production and/or livestock systems. These crop and livestock systems interact with the soil (i.e., which is part of the local biophysical system) and with pests and diseases. Therefore, although the biophysical environment provides the necessary condition for the presence of a farming system, it is the socioeconomic environment that provides the sufficient condition for its presence [Norman et al, 1982]. The current farming systems reflect the cumulative interaction of the socioeconomic and biophysical environments over time. Also what will exist in the future will be determined by their characteristics and interactions during current and future time periods.

The systems framework in Figure 2 implies that the composition, productivity, and sustainability of the farming systems are heavily influenced by factors beyond the farm gate. Characteristics of the local and regional biophysical and socioeconomic environment can be critically important influences. For example, characteristics of the local biophysical environment such as soil type, steepness of slope, rainfall level, and/or intensity, will be important in determining the level of, and potential for, soil erosion. However, the regional socioeconomic environment could also be very important in influencing the current and potential levels of soil erosion. For example, national policies on input subsidies and output prices will be reflected in local and farm gate prices for both purchased inputs and produce sales. This will influence which farm enterprises are undertaken, and the level of management practised. Whereas some may be beneficial from a soil conservation point of view, others may provide short term benefits at the expense of long term sustainability.

There is abundant empirical evidence emerging, particularly in high income countries, that current agricultural production methods may lead to what are

Figure 2: SYSTEMS FRAMEWORK

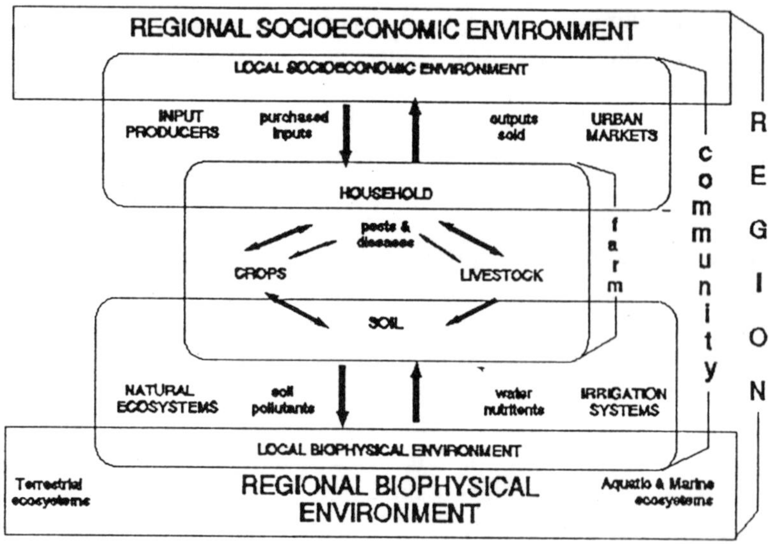

called social-trap problems [Williams, 1990].[16] For example in the USA, as Williams indicates, current agricultural policies attempting to maintain low food prices and high income levels encourage farmers to use fertilizers and pesticides to increase short-run production levels and profits. Obviously, such policies could easily result in soil erosion and degradation, on the fields

---

16 The term 'social-trap' "refers to situations in which an individual or society starts in a direction or relationship that later appears to be unpleasant or lethal, with no easy way to change or avoid the situation." [Williams, 1990: p. 28]. Such situations occur when there are conflicts between attractive short-run rewards for individuals and long-run costs for society.

of farmers using them and can also create problems for other producers and possibly the consumer (e.g., contamination of water supplies, chemical residues on food products, etc.). Unfortunately, farmers using large amounts of fertilizer may be deluded into a false sense of security. This is because its use can temporarily offset, to some extent, the effect of productivity losses caused by soil loss [Anderson and Thampapillai, 1990; van Vuuren, 1986; Dumsday, 1971].[17,18]

Therefore, an interdisciplinary approach obviously is necessary to develop land-use management strategies designed to provide attractive short-term benefits for current producers, while conserving the soil for productive use by future generations. This approach requires contributions from both the technical and social science areas. Before outlining the approach required, discussion in the remaining part of this section focuses on outlining, in general terms, the different strategies that can make a land-use system more sustainable. Hart and Sands [1991] suggest using the following strategies in identifying and analyzing potentially suitable alternative technologies:[19]

- ***Maintain or increase farm-by-biophysical environment interactions.*** Maintenance or enhancement of soil productivity is particularly important

---

17 Others have also argued that excessive use of fertilizer, particularly superphosphate, may damage the structure and overall quality of the soil [Costin and Coombs, 1981; Singh et al, 1987]. Also, some farmers in low income countries speak of the soil 'getting hooked' on fertiliser -- you have to keep applying them or the yields drop below what they were before fertiliser was used.

18 However, this would be an incorrect measure of what would have been the case without soil loss and fertilizer. Would mean yields be higher and variation in yields be less? The real question is what yields would be if soil erosion was reduced and new technologies increased yield, instead of masking the effect of soil loss.

19 These relate back to the material presented in Figure 1.

because there is a close link between soil productivity and plant production. More sustainable land-use systems can result from changing the ways in which the soil is managed. Management effects on soil productivity can be allocated to three different groups, physical, chemical, and biological. However, it is also possible to devise more sustainable land-use systems through using technology that positively affects soil/plant interactions. Hart and Sands [1991] propose that a possible reason why some cropping systems are not sustainable is that soil and plant management systems have been separated rather than integrated. Exploiting beneficial soil/plant cycles and decreasing nutrient losses that have to be replaced by external inputs can make the land-use system more sustainable. With respect to soil/plant relationships, the diverse cropping system usually found in more traditional agriculture obviously has merit. Systems consisting of cereals, legumes, and perennial plants, such as trees, can improve soil resource use efficiency, through tapping soil nutrients and water from different soil depths,[20] and with nitrogen-fixing species actually increasing soil nitrogen levels. This can help in improving the dependability of the system, thereby reducing farmer risk.

- ***Maintain or increase farm system component interactions.***
  Many different types of technology can be used to increase system sustainability on mixed farms, where both crop and livestock enterprises exist. Once again, exploiting the interactions between the various components in the system has merit. As Hart and Sands [1991, p. 5] indicate, the ideal system is one "where the 'waste' from one enterprise is an 'input' into a second enterprise." Obvious examples are:

---

20 Different species also vary in terms of the levels, composition, and times at which soil nutrients and water are required.

- o Using crop residues as livestock feed and bedding with the resulting manure being utilised for crops.

- o Agroforestry systems in which the tree component recycles soil nutrients from deep in the soil profile to crops and provides forage for livestock.

- o Decreased reliance on external pesticides through enhancing soil/plant/pest interactions (i.e., habitat management), etc.

- ***Maintain or increase farm-by-socioeconomic environment interactions***. These interactions are critically important in determining economic viability. Changes in policy can affect both economic viability in the short run and sustainability in the long run. Hart and Sands [1991] maintain that policy makers need to be provided with information that will help them predict the long-term impact of policies they propose to make. Policies that improve economic viability in the short run can also reduce farm-level resource productivity and have other detrimental external effects that reduce potential levels of livelihood of future generations. Policies should be implemented to ensure that this social trap does not exist.

Therefore, in view of the complex interactions that are possible, ensuring sustainability is a complex task -- something that farmers in low income countries have traditionally been very good at.[21] This is true not only because the net effect of changing the interactions is difficult to predict, but also because a large number of alternative combinations exist. Not surprisingly, Hart and Sands [1991] indicate that designing an alternative is

---

[21] That is, before population growth rates increased dramatically upsetting the relatively high land/labour ratios.

often more of an 'art' than a 'science'. This approach is at variance with the quest, until recently, for ever increasing levels of productivity in the short-run. This has been accomplished through using technology-intensive farming practices combined with specialized crops. These have been grown with extensive chemical additives manufactured with non-renewable resources, and at the expense of sustainability in the long-run [Napier, 1989; Francis and Hildebrand, 1989; Conway, 1987].[22] To create conditions that are more sustainable, new technologies must conform with the environments where they will be used, because most farmers are unable to modify their environments [Francis and Hildebrand, 1989]. One implication of this is that solutions designed to encourage sustainable agriculture will need to be based on a fuller understanding of the environment and on technology-by-environment interactions, implying that both the relevant solutions and evaluation criteria will be more location specific [Francis and Hildebrand, 1989; Hildebrand 1989]. A need exists to blend the traditional indigenous systems (i.e., often oriented to resource use conservation) with new technologies (i.e., often oriented to raising resource use intensity and productivity) to simultaneously ensure high use intensity and resource conservation [Jodha, 1990]. In other words there should be much greater emphasis on good land husbandry.

2.3 FARMING SYSTEMS DEVELOPMENT (FSD)

The methodology for the FSD approach provides some important elements that are necessary for the successful implementation of the new approach to achieving soil conservation. Detailed information concerning the characteristics, methodology, and implementation of the FSD approach are

22 Conway [1987] has also concluded that this approach has contributed to inequitable distribution of benefits in agriculture.

available elsewhere [FAO, 1989, 1990B]. Therefore, the following represents only a very brief summary.

The FSD approach developed by the Farm Management and Production Economics Service of FAO is based on the approach popularly known as FSR. The main objective of FSD is to improve farm-household systems and rural communities on a sustainable basis. It's immediate objectives include:

- Improving the efficiency in farm production.
- Raising farm and family income.
- Increasing welfare of farm families and satisfying basic needs.

Consequently, FSD is built on the development principles of improving productivity, increasing profitability, ensuring sustainability, and attaining equitability in terms of distribution of the fruits of production.

The farm-household is the principal system and focus of FSD and consists of three basic sub-systems, which are closely interlinked and interactive, namely, the household, the farm, and off-farm activities. The activities of FSD can be divided into the following two major categories, both of which involve substantial farmer participation (i.e., dialogue, interaction, consultation and, whenever relevant and possible direct 'hands-on' involvement):

- ***Farming systems analysis***

  - Viewing farm-household systems as a whole in order to understand connections and interactions between individual components.

  - Obtaining, with the help of farmers, an understanding of the biophysical and socioeconomic environments that influence decision processes at the farm-household level.

- Diagnosing constraints to increased farm production and productivity, as well as standard of living.

- Identifying and assessing, ***ex ante***, potential improvements.

- Where relevant, passing on the results, implications and possibly recommendations arising out of the farming systems analysis to involved parties. These could include researchers, extension and support service staff, and/or policy makers.

- ***Farming systems planning, monitoring, and evaluation***

  - Testing, monitoring, and evaluating improvements on-farm. Examples of such activities include those relating to proposed technological improvements, proposed revisions in farm plans, and improvements in support services and farm level impact of proposed policy changes.

  - Facilitating the spread of potentially successful technologies through extension.

  - Suggesting, where relevant, adjustments in the support services, research, and extension strategies and policy programmes that will facilitate technology dissemination and adoption.

  - Monitoring and evaluating the adoption rate of the proposed improvements by the farming community. This constitutes the major measure of success for the approach.[23]

---

[23] This sounds easy, but is in reality somewhat complicated and is likely to be frustrating if evaluated completely from an 'outside' perspective. Farmers are unlikely to initially adopt 'packages' of technology in their entirety. Rather, they are likely to adopt

We cannot sufficiently over-emphasise the importance of farmers' involvement in all activities -- something that has often been insufficiently stressed by many farming system practitioners. As will be seen later (i.e., Chapter 5), there has been, in recent years, rapid evolution of farmer participatory techniques (i.e., rapid rural appraisal (RRA) and participatory rural appraisal (PRA)) which make this interaction much more effective and productive.

A number of steps are designed to elicit information required in fulfilling the components of FSD as listed above. For convenience in presentation and to clarify thinking, these can be divided into four phases.[24] A very brief description of each phase follows:

- ***Description/Diagnosis***

  - After deciding on the location of the work, start-up activities involve reviewing secondary information on the area, making the necessary contacts, assembling the professional team, and making logistical arrangements.

  - Farming families are tentatively classified into homogeneous groups (i.e., sometimes called recommendation domains or target groups).[25]

---

specific components that fit well with their present systems. There can still be a very successful impact but perhaps not in the way 'outsiders' might view it! Whose views are more important -- obviously the farmers! We need to look at things through their eyes, not ours!

[24] Although Hart and Sands [1991] suggest five steps in addressing issues of sustainability, the overall approach is analogous to that of FSD.

[25] The FSD literature usually calls these farming system zones [FAO, 1989]. However, because the term 'zone' implies a contiguous area, it is not used in this publication.

The farming families within each group or domain usually practice the same farming systems, face the same constraints, and have the same potential solution(s) to their problems. This tentative classification can be modified as a result of information from the following two activities.

- An informal exploratory diagnosis using farmer participatory techniques, is undertaken to obtain a qualitative understanding of the determinants of the existing farming systems (i.e., both biophysical and socioeconomic). At the same time, efforts are made with farmers to ascertain the constraints, flexibility, and potential opportunities in the farming systems they are currently implementing.

- Sometimes a verification survey involving a structured formal survey, suitable for statistical analysis, is undertaken to quantitatively confirm insights obtained in the exploratory diagnosis. It can also be used to provide a database for farm and development planning, policy analysis, monitoring, and evaluation.

- ***Design***

  - In consultation with farmers, the constraints are ranked according to their severity, and potential solutions are identified after determining what flexibility exists in the farming systems currently practised.

  - Prior evaluation is made of the proposed solutions, which can be technological or institutional (eg. policy options) in nature. A number of analytical techniques can be used to evaluate the potential

---

Farmers in different recommendation domains can have both their residences and fields next to each other.

technical feasibility, economic viability, social acceptability, and ecological sustainability of the proposed solutions. One or more of the proposed solutions is selected for actual evaluation on-farm.

- ***Testing/Implementation***

  - On-farm testing or evaluation with farmers determines how well the potential improvements fit into the system, whether or not they are accepted by farming households, and what modifications are needed to make them acceptable. As indicated earlier, there are three types of on-farm evaluation, namely proposed technologies, proposed farm plans, and proposed changes in support systems and/or policies. These evaluations are undertaken with a small number of farming households, representative of the recommendation domains whose needs are being addressed.

  - Positive results of such evaluation provide justification for the FSD to advocate further action. The tested technologies can be disseminated through the extension service to other farming households in the same recommendation domains. The same applies to the farm plans that have been successfully tested. Favourable test results for proposed changes in support services and/or policies provide valuable farm-level information on necessary programme or policy adjustments (e.g., in extension, marketing of inputs and products, pricing policy, credit, etc.).

- ***Dissemination/Impact***

Monitoring and evaluating the rate of adoption of the changes earlier proposed, tested, and implemented provide insights into further desirable activities -- both by researchers and planners.

3. USING FSD FOR SOIL CONSERVATION INITIATIVES

3.1 INTRODUCTION

The objective of this chapter is to discuss in more detail some of the major components of the new approach to soil conservation and indicate their degree of compatibility with the FSD approach. Problems that are likely to arise in successful implementation of the new approach are indicated, and, where possible, suggestions are made for their resolution. The topics discussed include the following:

- Interactive linkages.
- Involving the farmer.
- Studying the farm-household system.
- Studying soil degradation.
- Short-term needs and long-run payoff.
- Preventing rather than curing soil erosion.

To provide a context for this discussion, the basic strategies of the new soil conservation approach are first outlined.

3.2 POSSIBLE SOIL CONSERVATION STRATEGIES

Discussion earlier in the publication implied that there are three relevant strategies for implementing the new soil conservation approach.

- ***Prevention thrust***. This involves preventing soil erosion from occurring through developing and disseminating land use enterprises, technologies and farm management practices that yield short-term production benefits while being conservation effective. This means they at least maintain (i.e., sustain) the productivity of the land for use in the long-term by future generations. The approach involves the following:

- o The selection of land use enterprises whose growth (i.e., physiological), management and conservation requirements match the characteristics of the land on which they will be undertaken.

- o Optimally the development and dissemination of conservation effective production technologies that not only have a positive production impact in the short run but also a positive conservation payoff in both the short and long run.

- o As a minimum, all potential technologies would receive ***ex ante*** screening to ensure, to the extent possible, that their adoption will have no negative environmental impact (i.e., they should at worst be conservation neutral rather than conservation negative).[26]

Because this thrust concentrates on promoting appropriate land uses and developing conservation effective technologies, it places major emphasis

[26] It should be possible during ***ex ante*** screening of proposed land use improvements to determine whether they are:

- ***Conservation effective*** -- the proposed improvement will be expected to directly or indirectly contribute to the maintenance and enhancement of soil productivity and prevent land degradation.
- ***Conservation neutral*** -- the proposed improvement will be expected to have no significant direct or indirect impact (i.e., beneficial or negative) on soil productivity or land degradation.
- ***Conservation negative*** -- the proposed improvement will be expected to directly or indirectly contribute to a decline in soil productivity and land degradation.

The prevention thrust requires that improvements aimed at yielding short term production benefits should at best be conservation effective, and at worst, conservation neutral.

on agronomic and biological means for ensuring sustainability. It thus concentrates on strategies designed to ensure good land husbandry.

- ***Policy thrust.*** This thrust can be important not only in preventing soil erosion arising in the first place but also in helping to stop it, once it has developed. It involves designing and implementing policies that:

  - Eliminate possible conflicts between policies designed to promote short-run production and those designed to encourage long-run sustainability (i.e., conservation).

  - Positively encourage, via appropriate incentives,[27] adoption of strategies that conserve the environment for use by future generations.

  The first approach is likely to be more important in preventing the development of soil erosion and in promoting good land-use management through the adoption of appropriate farm enterprises and technologies. The second approach, on the other hand, is likely to be more necessary in correcting erosion that has already developed.

- ***Corrective thrust.*** This thrust is somewhat analogous to the past conventional approach to soil conservation in placing emphasis on physical structures and other cross-slope barriers to prevent further

---

[27] The ideal incentive is tangible short term production benefits (e.g., higher crop yields, reduced input costs, increased production of fodder and fuel, etc.). The conventional approach has been to offer financial incentives such as cash payments, food for work or free farm inputs. If the Government revenue budget cannot sustain these payments post project, as is typically the case, then such incentives cannot be considered appropriate.

erosion, once it has developed. It may also involve the physical rehabilitation of severely degraded land by mechanical means (e.g., filling in of gullies, construction of check dams, etc.). The literature abounds with useful detailed information on various types of physical structures to combat erosion and in recent years this has expanded to include a range of vegetative techniques for controlling surface runoff and stabilising gullies. However, the new approach differs from the past approach in basing proposed solutions on an understanding of the causes of the soil erosion, rather than just on the symptoms.

The above breakdown of soil conservation strategies is simplistic because the three thrusts are not necessarily mutually exclusive. For example, policies can help or hinder the use of appropriate technologies to prevent soil erosion or the development of appropriate physical structures. Consequently, a combination of strategies is likely to be used to maximize the effectiveness of a soil conservation programme. However, the breakdown given does help to highlight some of the points made in the following sections.

3.3 INTERACTIVE LINKAGES

Two of the main strategies to improve the productivity, and hopefully welfare, of farming families on a sustainable basis are:

- The development of ***relevant improved technologies***[28] by research (i.e., with inputs in terms of feedback from extension and farmers) and their dissemination via extension.

---

[28] Improved technologies will include the development of new crop, livestock and tree enterprises that better match the characteristics of the land, as well as improving the field level management practices of existing farm enterprises.

- The development of ***relevant policies and support systems*** by planning and their implementation by extension, development programmes, etc.

Obviously, several groups of people or 'actors' are responsible for making this improvement in productivity possible in a sustainable manner (Table 1). FSD recognises that farmers play the most critical role in this endeavour. Not only are they ultimately responsible for actually implementing improvements in the productivity of their farming systems, but as we have already intimated, they can play critically important roles in helping to identify and develop the appropriate strategies. Researchers and planners are responsible for developing relevant improved technologies and the appropriate policy/support systems. Extension staff possibly with the help of other agencies, both in and outside the commercial/public sector disseminate advice, facilitate the distribution of improved inputs and help provide a market for the products to be sold.

For improved productivity to occur, close linkages are essential between farmers, planners, researchers (i.e., on-station and on-farm), and extension/development staff. The nature of these linkages is also important in determining the effectiveness of FSD. Until recently, the one-way, 'top-down' from researchers/planners via the transmitters to farmers was the norm. However, obviously what is optimal is for two-way interactive linkages to operate between all the 'actors' in the agricultural development process. FSD has facilitated the development of linkages from farmers. Where interactive linkages exist, FSD can thrive, and, as will be seen later, certain soil conservation strategies will then have a reasonable chance of being effective. Unfortunately, such effectiveness may be diminished where the link to the farmer (i.e., 'top-down') is still stronger than that from the farmer (i.e., 'bottom-up').

An example that illustrates the importance of linkages is implementation of the policy type of soil conservation strategy. Farm level work that demonstrates the need for changes in the policies or support systems will have little or no influence, if 'bottom-up' linkages do not function effectively. Currently, this tends to be the weakest link in FSD type work. Locating most of this work in technical research institutions has tended to encourage focusing on technological aspects. Indeed, policy/support systems often have been viewed as a given and not subject to change, as a result of farming systems activities. Fortunately, FSD explicitly rejects this submissive approach to the policy/support system, in favour of an approach that is more interventionist in character. Nevertheless, the ability to influence the policy/support systems as a result of farm level work has rarely been demonstrated.[29] Without creating an effective linkage with planners, the policy thrust cannot be effectively employed as part of a soil conservation strategy.

3.4 INVOLVING THE FARMER

Involving farmers has always been one of the basic tenets of the FSD approach. Part of the reason for advocating this 'bottom-up' method in the new soil conservation approach has been the lack of past success in getting farmers to adopt conservation practices. Technically correct conservation practices often fail because the socioeconomic circumstances of the farmers expected to implement them have been ignored [Douglas, 1990].

However, involvement of the farmer can also be stated in more positive terms. In FSD, two very obvious reasons why farmers (i.e., farming families) should be involved in the research process are recognized.

---

29 A recent FAO consultancy report addresses the issue of institutionalization of FSD [Norman and Pratt, 1990]. The organizational structure within which it is institutionalized influences what can be done.

- They have a right to participate. Because farmers stand to lose or gain the most from adopting changes proposed by researchers, extension, development agencies, and planners, they should have a 'voice' in developing a product that they want.[30]

- They are generally free to choose from the available technologies, and for that matter, generally the policy/support systems they wish to benefit from. Thus, if the technologies or policy/support systems being offered to farmers are not accepted by them (i.e., the ultimate consumers), all the work of the other groups delineated in Table 1 is wasted.

However, an even more fundamental reason for farmers' involvement in the process of developing relevant strategies is that they can improve the efficiency of the process. This may be either in terms of improving the probability of a return or reducing the quality and quantity of the inputs required to develop the strategies. Thus, farmers can contribute constructively during the:

- Descriptive/diagnostic phase, by articulating their needs or constraints, which can help in formulating the appropriate research and policy/support system agendas.

- Design phase, by helping to design and evaluate -- in an ***ex ante*** manner -- possible solutions to the constraints or needs they, and perhaps others, have articulated.

- Testing/implementation phase, by participating in the testing and ***ex post***

---

30 They, after all, often view problems, constraints, priorities, potentials and opportunities in ways that differ from 'outsiders' (e.g., technologists, planners, etc.). However, it is the farmers who have the last word!

evaluation of the potential solutions that were identified in the design stage.

The farmers' contribution to the different stages of the FSD process is important because of their intimate knowledge about:

- The biophysical environment (i.e., physical conditions) of their farms such as soils, water availability and movement, weed complexes, etc.

- The socioeconomic environment (i.e., socio-cultural and policy-institutional conditions) and the quantity and quality of the resources they have at their disposal, such as labour, land, capital (e.g., draught, equipment, etc), and managerial expertise.

- The complexity of their farming systems which results from allocating their limited inputs (i.e., resources) to cultivating a number of crops, to often keeping more than one type of livestock, and sometimes also to pursuing an off-farm occupation.

Many examples have been cited of how including farmers in the different stages of an FSD type process has benefitted the formal research process [Chambers et al, 1989]. In spite of this, farmer involvement is often only recognized as being desirable rather than being implemented in practice. Possible reasons for this include:

- A 'development for the farmer' approach is often used, which is merely a thinly disguised version of the old 'top-down' methodology. The basic assumptions behind this approach are that the farmer is ignorant and that the 'specialist' knows best. However, because the ecological, social, and economic factors in agriculture are complex and dynamic, only the land-user can make informed on-the-spot decisions [Bartlett, 1978].

- Farmers are often considered to be passive and conservative. Actually, however, the majority experiment continuously. They often combine this with a very accurate practical knowledge of their environment, which is usually much more subtle than an outside specialist could develop. They are also invariably more capable of recognizing the relationships between various factors influencing their environment. This enables them to have a more holistic view of their problems compared with the narrower perceptions likely to be possessed by external specialists trained in one discipline [Chambers, 1983; CDCS, 1986; Richards, 1985].

- There is a lack of skill or knowledge on how to fully involve the farmers in the research/development process. Fortunately, as hinted at earlier, this situation has changed greatly in recent years with many references becoming available, based on practical experience. Much of this work is presented under the farmer participatory research rubric [Chambers et al, 1989; Farrington and Martin, 1988; Haverkoort et al, 1991].[31]

Hence, the approach should be based on ***development with the farmers*** and, ultimately, should lead to ***development by the farmers*** [Rhoades and Booth, 1982]. Thus, farmers, become active participants in the development of appropriate conservation effective farming systems and the role of the technical specialist is one of ***working with farmers for better land husbandry*** [Hudson and Cheatle, 1993]. However, involving farmers means that they can not only contribute but can also criticize, change, or resist any activity. This is not always easy for external 'experts' to accept!

---

[31] In Botswana, the use of self-selecting farmer groups, which meet to discuss issues but are individually responsible for what they do, has been found to be an effective forum for involving farmers and creating a collegiate and interactive relationship with researchers [Heinrich, Masikara et al, 1990].

TABLE 1: ROLES AND FUNCTIONS OF THE 'ACTORS' IN AGRICULTURAL DEVELOPMENT

ROLE	FUNCTIONS	ACTORS
Implementing	Implementors	Farmers, Communities
Supporting	Transmitters, Input Provision, and Market for Products	Extension Staff, Development Agencies, Non-Governmental Organisations, Commercial Firms
Providing Potential Means	Technology Policy/Support Systems	Research Planning

For farmers to be able to make constructive and significant contributions to the research process, it is critically important that good relationships develop between them and the other groups listed in Table 1. The keys to this are mutual respect and the development of a collegiate or partnership type of relationship between the external staff and farmers. Often, a paternalistic type of relationship exists because of differences in educational levels. If this is the case, then the development of a constructive form of collaboration will be difficult. This is because in a paternalistic relationship, for example, only those technologies that researchers feel confident about are likely to be offered for testing by farmers. However, if a partnership develops, then an attitude of joint investigation can flourish, allowing any of the technologies to be tested with farmers. Under this partnership -- unlike the paternalistic type of relationship -- blame is unlikely to be apportioned to the external staff, if the trial fails. Other elements contributing to developing a good relationship include the external staff being: sensitive to the local customs; respectful; friendly; willing to listen; and patient. They also need to make sure all parties understand and accept the potential obligations and benefits of any proposed collaboration.

Farmers, as suggested earlier, do not form one homogeneous group. Even at the village level, there are many differences between farmers, for example, in family size, wealth, power, age, gender, social status, access to markets, character, experience, etc. Thus, there may be important differences in the felt needs, perceptions, and suitable strategies to help different types of farmers. For example: women often have very different needs from men; richer and more powerful land users are often more conservative than poorer ones; and market-oriented farmers, even when poor, have different needs and priorities than subsistence farmers [Chambers, 1983]. Yet, such differences have rarely been recognized in soil conservation programmes. Thus, as discussed earlier, it will be necessary to divide the farming community into homogeneous sub-groups or recommendation domains to ensure that the same land husbandry or soil conservation strategy is relevant for all the farmers in that domain. However, what is relevant for one group of farmers will often not be relevant to another group, so different land husbandry or soil conservation strategies may be needed for farmers in different recommendation domains.

This heterogeneity in the farming community has another important implication. The successful implementation of some soil conservation strategies (e.g., particularly the corrective thrust) requires cooperation between a number of farmers. Obvious examples are terracing on steep slopes, structures to combat gully erosion, and strategies that optimize watershed management. Heterogeneity in the farming families that make up the land-user groups, which are necessary to implement such strategies, can cause problems. This is particularly likely to arise if there has been no history of cooperation for this or other purposes. Generally, unlike those involved in applying the more traditional soil conservation approach, farming systems practitioners have had little experience in the dynamics of forming land-user groups that require collective responsibility and action.[32]

[32] Possible exceptions in recent years have been the experiences gained by some practitioners working in the agroforestry area [Raintree, 1987; Muller and Scherr, 1990].

However, intrinsically, the fundamental notion of involving farmers in the various FSD activities should provide an appropriate forum in which such groups can be planned, implemented, and monitored.[33]

3.5 STUDYING THE FARM-HOUSEHOLD SYSTEM

The vast majority of households living in rural areas in low income countries are [Ellis, 1988, p. 12]:

> "Farm households, with access to their means of livelihood in land, utilizing mainly labour in farm production, always located in a larger economic system, but fundamentally characterized by partial engagement in markets which tend to function with a high degree of imperfection."

A number of microeconomic theories for farm households have been developed in an effort to explain their economic behaviour. These are based on various assumptions about household goals and the characteristics of the markets within which households make their decisions. Ellis [1988] suggests that the theories predict different results because of different assumptions about the working of the factor (i.e., land, labour, and capital) and product markets, rather than because of differences in assumptions about household goal(s). In fact, an important characteristic distinguishing many of the theories results from different assumptions about labour markets and the

---

[33] Such community participation may, however, not always be necessary. For example, greater emphasis is now being placed on simple vegetative techniques involving building up terraces by natural means over time rather than solving the problem through constructing physical structures. The former, for example, may be achieved through the slow build up of soil on the uphill side of grass strips.

allocation of household labour time. Consequently, a certain universality is likely to exist in the aspirations of farming households. The most important elements of these aspirations are income, effort avoidance, and risk avoidance. In economic terms, this means that such households try to increase their utility (i.e., satisfaction), which increases with income but decreases with greater effort or higher levels of risk. This can be restated as maximizing income for any given level of effort or risk. Alternatively, it can mean reducing risk, or effort, for a given level of income [Norman, 1982].

Attempts to maximize satisfaction take place within a set of constraints. It is probably variation in the constraints, rather than the aspirations, that leads to the most important differences in farming systems. It is easy to appreciate constraints relating to the biophysical environment causing differences in the type and productivity of farming systems. However, also important in the differentiation process are the factors relating to the socioeconomic environment. The quantity, quality, and ratios of the factors of production (i.e., particularly between land and labour and also capital) play an important role in further differentiating farming systems. The extent to which they can or will be modified will depend on the degree of integration into the factor markets (i.e., input markets such as hiring labour). Also, the products resulting will be influenced to some extent by the socio-cultural environment (e.g., religious taboos may prohibit tobacco growing) and the degree of integration into the product market (i.e., what and how much they are willing to sell). Greater integration into the factor and product markets would imply that market influences and government policy/support systems would play more significant roles. As will be discussed later, an understanding of such relationships and constraints is very important in helping design appropriate technologies that will be attractive to farming households. They will also determine what will be the most appropriate land husbandry or soil conservation strategies.

The word 'system' in the term 'farm-household system' indicates that the three components (i.e., the farm, the household, and off-farm activities by household members) are related to, and interact with, each other. Two reasons why it is important to stress the link between the farm-household system and land degradation, are as follows:

- Although often not emphasized enough in the past, land degradation is obviously strongly related to the farm-household system. Land degradation becomes an important issue because of its negative impact on land productivity and, eventually, human welfare.

- Land degradation often manifests itself on the farm and is usually caused by land management practices undertaken by household members in fulfilling household needs.

Consequently, to treat the causes -- and not just the symptoms -- of land degradation, the farming system and, more specifically, the household needs and household decisions, have to be analyzed. Thus, soil conservation activities have to be integrated into productive farming system activities rather than being offered as separate land management activities [Douglas, 1990].

FSD principles can be used in accomplishing this. Once recommendation domains have been established, farm-household systems representative of each domain can be studied in greater detail with the aim of deriving suitable strategies for soil conservation [Douglas, 1992]. Farm-household information helpful in the process of drawing up suitable strategies with the farmers themselves can include:

- An analysis of household and village organization.
- A description of the biophysical and socioeconomic environment of the household.

- A description of actual land use, including farmers' knowledge and views on soils, climate, crops and cropping practices, and constraints.
- A description of the household economy, including needs and production in kind and cash.

3.6 STUDYING THE PROBLEMS AND CAUSES OF SOIL DEGRADATION

Conventional soil conservation programmes have concentrated on trying to arrest soil movement. Loss of soil was considered a problem for:

- The farmer -- in the form of productivity loss or loss of land.
- The government -- in the form of loss of natural resources and damage to infrastructure.
- Sometimes for other land users -- for example, sedimentation of irrigation schemes or downstream flooding problems.

However, loss of soil is only a symptom of soil degradation. Symptoms have causes (e.g., faulty policies favouring poor land management) and effects (e.g., loss of productivity). By concentrating only on soil loss, past soil conservation programmes neglected both its causes and its effects. They automatically assumed that soil loss was creating problems. Whether this was actually true, and for whom this was true, were almost never verified. As implied earlier, they assumed that treating the symptoms would be sufficient to cure the problem.

Ignoring the effects and causes of soil loss in the past led to programmes adopting a 'top-down' physical planning approach. Symptoms of soil degradation are generally visible to a specialist, even on casual inspection. Also, the related degradation processes, such as soil erosion, can be quantified relatively easily in terms of topography, soil, and climate.

However, it is much more difficult to define whether the visible signs of erosion are actually causing problems, and if they are, for whom and to what extent. To determine this, information is necessary about the production systems, about the changes that have taken place in them over time and why they have changed. These changes could be physical, cultural, or socioeconomic in nature and could contribute directly or indirectly to the soil degradation problem.

Therefore, substantial amounts of information are required, if the aim is to have a soil conservation programme that decreases soil degradation problems rather than soil degradation symptoms. Often, soil erosion originates from specific land use and land management practices. It is generally relatively easy to identify which practices are responsible for the degradation. However, it is not always so easy to determine why these practices are carried out and how these might be changed to arrest soil erosion in the future thereby encouraging sustainability. For instance, overgrazing, deforestation, or ploughing up-and-down the slope can often be quickly identified as causes of erosion. Nevertheless, ascertaining why land users follow these practices, and what needs to be done to get them to change is a more complex task that demands a thorough knowledge of land users and their production systems.

Consequently, any conservation programme aimed at solving the degradation problem through treating its causes necessitates a 'bottom-up' programme, based on detailed knowledge of the production system and the effect soil degradation has on that system. However, shortcuts can be made in the amounts of data that are collected and the types of analysis that are necessary. For example, land suitability and capability information[34] for

[34] Land suitability indicates the best way to use the land to maintain productivity, whereas land capability ranks land according to its relative erosion hazards [Shaxson et al,

different soils can be very useful in giving insights into soil characteristics and quick guidelines on the best type of management.

3.7 SHORT-TERM NEEDS AND LONG-RUN PAYOFF

Because farming households are the primary clients for FSD type work, FSD largely responds to the 'felt' needs articulated by them. The closer farmers are to the survival level, the greater will be the likelihood that felt needs will be those that require fulfilment in the short-term (e.g., producing enough food to survive until next year). As a result, they are unlikely to be interested in societal concerns about environmental degradation in the long-run. Davies et al [1991] and Frankenberger and Hutchinson [1991] have stressed the critical nature of this relationship between food security and sustainable resource management. If there is a conflict between strategies required for sustainable resource management and those ensuring food security in the short run, problems will arise. Thus, strategies required to ensure food security must receive priority, if schemes guaranteeing sustainable resource management are to be successfully implemented. Only relatively rarely, where soil erosion problems are very severe and threaten immediate survival, will farmers be inclined to implement soil conservation measures spontaneously.[35]

Consequently, most farming systems type activities have tended to have a short-run focus. This is in contrast to the longer time frame required for soil

1989]. Land suitability evaluation [FAO, 1976] in particular, is believed to have considerable potential as a methodology for assessing the suitability of current and potential land uses at the local level.

35 See, for example, the work of Fujisaka [1989] in the Philippines.

degradation problems to become apparent. The FSD approach does stress the use of evaluating any proposed changes in terms of sustainability criteria, but this may be easier said than done, unless it is inherently obvious that a conservation effective production technology is being proposed (e.g., incorporating both cereals and legumes in a proposed system can have both short and long term production benefits) (see Sections 2.2. and 3.8).

However, this is not so easy in the case of technologies that do not obviously contribute to sustainability as well as productivity (i.e., do not have an obvious positive impact on the biophysical environment. The methodological complexity of incorporating societal evaluation criteria (e.g., sustainability and/or degree of equitability in the distribution of benefits) and the time required in the ***ex post*** derivation of such societal impact evaluations have severely limited farming system type activities in this area. Practically, emphasis probably will continue to be placed on rather subjective ***ex ante*** evaluations. However, such evaluations can potentially influence the solutions to be advocated from solving the 'short-run' felt needs of farmers. Over time, however, societal evaluations can be made more tangible, through monitoring what happens during and after the adoption process. This information can be useful in confirming the validity of the existing ***ex ante*** evaluation criteria. Alternatively, they may help in developing other evaluative criteria. Therefore, it is possible to develop technologies that do not have a negative societal environmental impact.[36] Of course, it is also possible to develop technologies that have a positive impact on the biophysical environment (see Sections 2.2 and 3.8).

In the absence of other pressures or incentives, farming families are free to adopt or not to adopt technologies that are thought not to be detrimental to

---

[36] However, this is likely to be potentially easier in the case of the biophysical environment (e.g., soils) than it is for the socioeconomic environment (e.g., with respect to the equitable distribution of benefits).

the environment. As emphasized earlier, some short-term benefits as a result of adoption usually will need to be obtained. Issues of both ability and disposition to adopt proposed improved technologies enter into the decision making process of farming households:[37]

- Ability to adopt is determined by such factors as to whether the technology is:
    - Technically feasible (i.e., whether the necessary inputs -- both on and off-farm -- are readily available).
    - Economically viable (i.e., whether the technology is profitable and dependable at the levels farmers are likely to adopt).
    - Socially acceptable.
- Disposition to adopt is decided partly by whether farmers will be better able to achieve their goal(s) by using the improved technology. Criteria on which they will decide their willingness to adopt are often difficult to determine.

Thus, even evaluating whether a technology with short-term benefits is appropriate for specific farming households is a daunting and complicated task. For example, the levels, qualities, and relative proportions of the factors of production (i.e., land, labour and capital) are important in indicating the route to follow in improving the productivity of farming systems, at least in the short-run [Hayami and Ruttan, 1971; Binswanger and Ruttan, 1978]. This route is appropriate if profit maximization is the aim. Attaining this involves maximizing the return to the most limiting factor. In

---

[37] Douglas [1992] also discusses the criteria in some detail including considerations such as whether it is: technically possible, practically feasible, productive, financially desirable, stable, sustainable, universally applicable, and socially and economically acceptable.

areas of high land/labour ratios (i.e., low population density) labour saving types of technology (e.g., mechanical tillage, use of herbicides) are likely to be most appropriate. In contrast, in areas of low land/labour ratios (i.e., high population density), land saving technologies that maximise production per unit area (e.g., intensive intercropping, agroforestry systems, integrated pest management, and some use of chemical fertilizers) will likely be more appropriate.

However, as Table 2 shows, increasing the return to the most limiting factor can have an indirect positive or negative impact on the use and productivity of other inputs. Thus, in view of the imperfect operation of the factor markets, care has to be taken in determining what would be the most relevant technologies. For example, initial superficial examination indicates that because of the high population densities, strategies designed to increase the productivity of land would be most important. However, the more seasonal the agricultural production cycle is, the greater will be the seasonal variability in the demand and use of labour. This will create labour bottleneck periods, followed by periods of underemployment. Thus, land-augmenting strategies using more labour at critical labour bottleneck periods -- when the opportunity cost of labour is high -- could be unattractive to the household. This perception will also be likely if the productivity per unit of labour applied at that time is higher in alternative uses. Additional complications may be introduced in dealing with this problem, for example, if restrictions exist in substitutability of male and female household labour. Inclusion of intra-household relationships or gender issues adds another complication. Not surprisingly, the complexities of accurately evaluating technologies in terms meaningful to farming households has been a major reason for including farmers in the designing, testing, and evaluation of technologies in the FSD process.

Turning to the issue of long-term benefits, the preventative type of soil conservation can play a benign or enhancing role in ensuring ecological sustainability. However, the policy thrust relating to soil conservation also can be very influential in encouraging or discouraging the adoption of technologies developed under the preventative thrust. Heavily subsidized chemical inputs, for example, may discourage the adoption of technologies that rely less on external chemical inputs and have a neutral/positive rather than negative impact on ecological sustainability.[38]

TABLE 2: RELATIONSHIP BETWEEN TYPES OF REQUIRED TECHNOLOGY AND LAND/LABOUR RATIOS

LAND/LABOUR RATIO	TECHNOLOGY REQUIRED	PRODUCTIVITY OF[a]	
		LAND	LABOUR
High	Labour saving	I+ or I-	D+
Low	Yield increasing	D+	I+ or I-

a. D = Direct impact I = Indirect impact
\+ = Positive impact - = Negative impact

The policy thrust can be used in two other ways to encourage farmers to adopt strategies encouraging soil conservation. These are:

- ***By subsidizing implementation of soil conservation strategies to make them more palatable to farming households.***
  This approach is likely to have most applicability in encouraging the adoption of soil conservation strategies relating to the corrective thrust,

[38] For example, in former West Germany, significant reductions in herbicide use have resulted from using economic threshold criteria in regulating weed control programmes for small grains [Walters et al, 1990].

particularly should there be a need for physical structures, or where farmers are required to forego short term production to secure future sustainability (e.g, switching from growing annual food crops to perennial crops that may not give any return in the first few years). It is not likely to appeal to farming households that are close to the survival level and who do not perceive soil degradation/erosion as being an immediate threat. After all, soil itself is not a benefit. Rather, it leads to benefits, such as production. Similarly, soil conservation itself offers no immediate tangible benefits. Instead, it may lead to benefits in the future. Finally, and very importantly, the need for subsidies to support this type of policy thrust will likely limit its use in many low income countries. Severe governmental budgetary limitations are likely to preclude much subsidized support being given to strategies that have a payoff sometime in the future.

A possible exception to this is, however, the constructive use of donor food aid to help those farming families whose food security is threatened. As indicated earlier in this section the process of maintaining household viability may occur at the expense of the physical environment. After farmers have exhausted the strategies that are feasible early in a food crisis (e.g., reducing consumption levels, changing cropping patterns, selling assets including small livestock and household valuables), they then start selling assets that reduce their productivity in the future. They also engage in practices that accelerate degradation of the physical environment (e.g., trying to grow crops on very marginal land, etc.) [Watts, 1988; Frankenberger and Goldstein, 1990]. Frankenberger and Goldstein [1992] have suggested that awareness of these so-called local coping strategies, combined with monitoring of the local conditions and trends (e.g., asset exhaustion, store depletion, migration, etc.), can help identify areas where immediate interventions are required. They have been instrumental in helping develop methodologies for assessing these in a rapid and efficient manner.[39] When these areas/families are identified, food aid can then be targeted in such a way as to improve use and management of resources.[40] With such an approach, food for work programmes can be implemented to promote soil conservation, water harvesting, promote agroforestry, reestablish forest reserves and other strategies that will restore degraded resources and improve agricultural productive potential can also be undertaken.[41] There are many countries where such food aid has enabled food for work programmes to be implemented with the objective of restoring degraded resources and improving agricultural potential. Typically such programmes have promoted soil conservation, water harvesting, agroforestry, afforestation and other field level activities, by paying farmers for their labour with food. Such programmes through aiding families during food crisis, help

---

39 They suggest this can be done through monitoring three types of indicators based on the work of the World Food Programme. These are: ***leading*** or ***early*** (e.g., crop failures, rapid decrease in condition of range and livestock); ***concurrent*** or ***stress*** indicators (e.g., migration, decreased consumption, selling of assets, etc.); and ***trailing*** or ***late outcome*** indicators (e.g., high rates of morbidity and mortality, land sales, permanent migration, etc.). Obviously interventions would be desirable as soon as the early indicators become obvious.

40 There is, however, a very delicate balance between supporting such families through crisis periods versus generating a dependency syndrome in the future.

41 An argument against using direct incentives, such as food aid, is that a decreased sense of involvement and responsibility can lead to poor standards of work and to poor maintenance [Hudson, 1991]. Farmers may regard themselves as hired labourers rather than participants [IFAD, 1992; Prior, 1992]. In one Ethiopian project farmers do not refer to 'food for work', but to 'work for food'. The food is the most important thing, the nature of the work is secondary. They also feel they are paid to build bunds not to maintain them. In some instances farmers have knocked down the bunds, hoping to be paid more food for rebuilding them the following year [Hilsum, 1987].

decrease the necessity of selling productive assets which would decrease future productivity [Frankenberger and Goldstein, 1992]. However it should be recognized that using food aid as an incentive for soil conservation may not always achieve the long term desired end result.

- ***By linking policies designed to encourage production with those designed to encourage conservation.***
  This 'carrot and stick' approach [Napier, 1987] fosters the notion that if something is taken out of the land to encourage production, something else needs to be put in to sustain land productivity in the future. This approach requires that the farmers participate in a specific conservation practice, if they are to benefit from programmes designed to stimulate production. For example, two development programmes in Botswana have provided incentives for destumping to improve the efficiency of the ploughing operation. This could have a negative impact by encouraging erosion, thus, lowering the potential productivity of the land in the future. A constructive policy would be to encourage destumping along with a programme of planting windbreaks, living hedges, etc.

  In the USA in recent years, production and conservation policies have been increasingly linked. For example, government subsidized loans may only be available to purchase inputs if the farmer pursues certain conservation strategies. These are known as 'cross-compliance policies.' To date, such policies appear to have been absent in low-income countries [Anderson and Thampapillai, 1990]. However, such cross compliance policies may be very difficult to put in place for farmers operating near the survival level (i.e., food security is the major over riding issue). Such policies are likely to be more effective in the case of farmers who have more resources (i.e., attaining food security is no longer an annual problem) and have a more commercialized orientation [Frankenberger, 1991].

Market prices for farm produced products may also have a major effect on farmers' enthusiasm and the production practices they adopt. However, the structural adjustment process currently under way in many countries, and the elimination of cheap food polices for the urban populace, should have a beneficial effect for the farmers who produce a surplus of food for the market.

The major objective of the strategies just discussed is to bring about a convergence between private short-term interests of farming families concerned about attaining an adequate standard of living now and the long-term societal interest in maintaining the environment for future generations. Given the realities that currently exist in many low income countries, the 'carrot and stick' strategy is likely to be more practical than encouraging adoption of conservation strategies through direct soil conservation subsidies. Obviously, whenever possible, the preventative thrust (i.e., improved land husbandry) provides the least costly and most attractive approach.

3.8 PREVENTING RATHER THAN CURING SOIL EROSION THROUGH IMPROVED LAND HUSBANDRY

Past conservation programmes have been based on the assumption that runoff is the chief cause of erosion losses, and that runoff and erosion are an almost inevitable consequence of using land for crop production [Shaxson 1989]. The problem has been perceived as one of how to control runoff on agricultural lands and in so doing to stop soil erosion. Past conservation projects have therefore, not surprisingly, adopted a 'top-down' physical planning approach, relying heavily on the construction of conservation banks, storm drains, artificial waterways etc., to control runoff and prevent soil movement.

The new approach requires a change of development focus away from soil conservation ***per se*** to what has been termed land husbandry [Hudson 1992]. The concept of husbandry is widely understood when applied to crops and animals. As a concept signifying understanding, management and improvement, it is equally applicable to land [Shaxson et al, 1989]. The primary objective of land management should therefore be improved, sustainable production through good land husbandry with control of soil erosion following as a consequence. This is a reversal of the previous idea that it is necessary to conserve the soil in order to get better crops [Shaxson et al, 1989].

A recent workshop in East Africa [Hudson and Cheatle, 1993] focused on two related concepts, one the need to integrate conservation practices into farming systems, the other the need for a participatory approach to conservation farming. The conference concluded that conservation farming can be achieved by a development approach that leads to professional agents ***working with farmers for better land husbandry*** [Cheatle, 1993]. The workshop arrived at the following tentative definition of land husbandry:

> "In the context of small-scale farming, land husbandry involves land users making decisions to manage their resources of soil, climate, plants, animals, implements, labour, knowledge and capital. Improvements to land use, and management practices should develop from the knowledge, problem awareness, analysis, ideas, insights, goals, aspirations and priorities of the farmers and their households. Decisions about changes to existing land use practices should involve fullest participation of the land users in identifying problems and opportunities, formulating and implementing appropriate courses of action and in monitoring and evaluating the results of doing so."

The above recognises the need for a participatory approach to conservation farming to enable sustainable management of natural resources by rural people. This definition effectively draws conservation into the mainstream of small-scale farmer development, by emphasizing improved land use and linking the concept of land husbandry with people's participation [Cheatle, 1993].

3.9 SUSTAINABILITY CRITERIA

As has been repeatedly emphasized, the fundamental principle of the new approach to soil conservation is that prevention of soil erosion is much more desirable than trying to treat it after it has happened. Therefore, emphasis is currently increasing on including sustainability evaluation criteria when assessing the appropriateness of proposed technologies. However, as also discussed earlier, sustainability is methodologically not a particularly easy evaluation criterion to assess in an *ex ante* manner or in a reasonable time frame prior to technology dissemination. Indeed, considerable controversy continues regarding selection of simple measures that can satisfactorily assess sustainability. With methodologies rapidly evolving in work on sustainability, perhaps simple, generally acceptable, direct measurements or proxy variables will soon become available.[42]

In the meantime, a more indirect approach to ensuring sustainability is emerging. Thinking is advancing on what types of strategies are likely not

---

[42] Harrington [1991] has reviewed some of the methods that have been used. These have usually been expressed in yield or production terms. However, as he indicates, such data (i.e., particularly when aggregated) can be misleading and mask changes in input levels or in enterprises allocated to different land types. The best indicator of sustainability will undoubtedly be one that measures the intrinsic 'quality' of the soil over time. Unfortunately, there is no such generally accepted measure, such as percent organic matter, changes in soil structure, etc. In any case, farmers are unlikely to be impressed by such measures. Their evaluation is likely to be based on what that asset can produce in terms of output.

only to maintain but also to enhance sustainability. Earlier, mention was made of Hart and Sands [1991] proposing that this can best be achieved through increasing interaction between the different parts of the farm system and between the farm and the environment (i.e., both biophysical and socioeconomic). In the same vein, a better understanding of the biophysical environment can help determine the categories of technology that are likely to contribute to sustainability. Possible examples include the following [Francis and Hildebrand, 1989; Conway, 1987]:

- Diversity -- mixed cropping, crops and livestock.
- Balancing/recycling -- rotations, alley cropping, agroforestry.
- Biological or cultural control methods -- integrated pest and habitat management.
- Tillage -- reduced through mulches, minimum and zero tillage.

The list indicates a heavy emphasis on biological measures,[43] supplemented to some extent by physical measures, mainly mechanical in nature.

Shaxson [1988] has well summarized the issues relating to sustainability and control of soil erosion (Figure 3). The types and sequences of plants that farmers grow are determined to a large extent by profit and/or subsistence, whereas those preferred by conservationists will be determined by a concern for soil conservation and water quality. Plants that can be used to achieve these two objectives may, or may not, be different. Activities improving soil conditions for rooting in essence favour the farmer's aim of more vegetative

[43]. This in itself is a major challenge in that some kind of balance is generally required. For example, livestock produce manure. Too much manure can sometimes cause more problems than commercial nitrogen, while success with minimum or zero tillage often requires heavy use of chemicals -- at least in high income countries. However, the quality of 'kraal manure' (e.g., in Africa) is usually such that this will not pose an immediate problem.

production per hectare. In addition, good crop cover and plant residues that assist the farmer's goal of good infiltration capacity and soil structure also help fulfil the conservationist's aim. Therefore, Shaxson [1988] suggests that, in order to ensure sustainability or to solve any problem of erosion by water, it is critically important to maintain appropriate overall combinations of the following key features:

- Effective depth for rooting.
- Structural soil conditions at this effective depth.
- Duration, density, frequency, and above-ground location of plant and residue cover.
- Steepness of slope.
- Length of slope.

As Shaxson [1988] notes, slope steepness and soil depth are relatively unchangeable, but slope length can be changed. However, the most critical variables under the control of the farmer are crop cover and soil structure. Finally, Shaxson [1988] emphasizes that negatively changing one of the above factors without making compensatory changes in one or more of the others usually creates an unstable situation.

FIGURE 3: INTEGRATING WATER AND SOIL CONSERVATION WITH AGRICULTURE

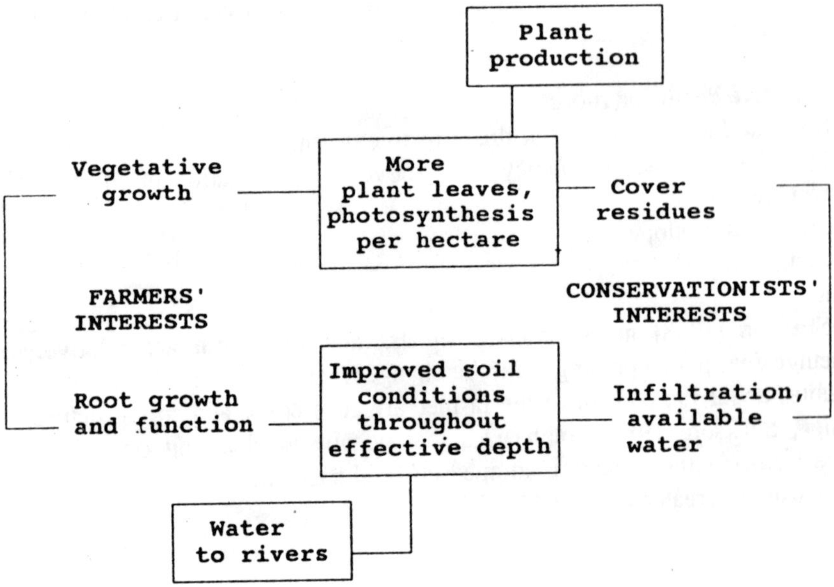

Source: Shaxson [1988]

4. IMPLEMENTING THE NEW SOIL CONSERVATION APPROACH: THE CONCEPTUAL FRAMEWORK

4.1 INTRODUCTION

In this chapter we outline in broad terms the major components of the conceptual framework for implementing the new approach to soil conservation. In the next chapter we outline in more concrete step-by-step terms what needs to be done in enabling these components to be addressed. The application of FSD methodology to implementing the new approach to soil conservation is still relatively untested in field situations. Therefore, the material presented should be considered more indicative rather than definitive in nature. Experience in implementing the approach in field situations will undoubtedly lead to adjustments and refinements in the approach and techniques. However, based on past success in applying the FSD type methodology to address the 'felt' needs of farmers, there is no doubt that the basic principles and steps have the potential of being successfully used in implementing the new soil conservation approach.

The lack of a single, detailed, definitive approach arises as a result of two other considerations.

- Some locational specificity is likely to exist in terms of what needs to be done in order to implement relevant strategies. For example, little additional effort will be necessary in gleaning additional knowledge and designing/testing solutions, if information relevant to the issues being considered is already available in the particular area.[44] In addition, of course, need for such information from 'outside' sources is reduced if the knowledge possessed by farmers' themselves is obtained and fully exploited.

- The approach and information required, to some extent, will be dependent on the type or types of soil conservation thrusts (i.e., preventative, policy, and/or corrective) that are deemed most appropriate.

Bearing the above caveats in mind, the approach required in applying the new soil conservation approach is outlined in Table 3, which is modelled on the format established for FSD. However, not all components are relevant for the different thrusts. Therefore, a tentative indication of the components required in applying the preventative and corrective thrusts is given in Table 3. The policy thrust has deliberately been omitted, because, by itself, it usually will not constitute a soil conservation strategy.[45] Instead, it generally plays an important facilitative role in supporting the effective implementation of the preventative and corrective thrusts. This is why a component is included for assessing whether there is a need to make policy adjustments.

In the following sections, the approach that needs to be used in addressing the components is outlined in general terms.

---

[44] For example, good information on the land suitability and capability of the different soils in the area would help reduce the amount of time spent on descriptive/diagnostic work.

[45] One possible exception would be considering the implementation of cross-compliance policies.

TABLE 3: SUMMARY OF THE STEPS NECESSARY IN APPLYING THE NEW SOIL CONSERVATION APPROACH

STAGE	STEP	STEP NECESSARY IN	
		PREVENTATIVE THRUST	CORRECTIVE THRUST
Description/ Diagnosis			
	Start-up activities	Yes	Yes
	Definition of recommendation domains	Yes	Yes
	Description of soil degradation	No[a]	Yes
	Assessment of causes, effects and risk of degradation	No[a]	Yes
	Ranking of constraints	Yes	Yes
Design			
	Preliminary selection of potentially suitable solutions	Yes	Yes
	Ex ante evaluation of proposed solutions	Yes	Yes
	Assessment of need for policy adjustments	Yes	Yes
	Final selection of solutions for testing/implementation	Yes	Yes
Testing/Implementation			
	Testing	Yes	No[b]
	Implementation	Yes	Yes
Dissemination Impact[c]			
	Degree of adoption	Yes	Yes
	Proposed modifications	Yes	Yes

a Since soil degradation has not occurred, there is no ***ex post*** analysis to be done. However, there is still value in knowing the potential for erosion to occur and using it as an input in the ***ex ante*** evaluation of proposed solutions (i.e., in the design stage), to problems that have been detected.

b Because the more conventional soil conservation approach placed considerable emphasis on physical structures it is unlikely much testing will really be necessary before full-scale implementation. In any case, it is simply unlikely to be feasible to do much small-scale testing of some types of physical structure since they would require full-scale implementation for proper testing!

c This is likely to be very dependent on the extent to which farmers themselves have been involved in all the earlier stages.

4.2 DESCRIPTION/DIAGNOSIS

4.2.1 Start-Up Activities

Initially a target area or general location needs to be selected. Often this will be a political decision and probably should remain so. Technical staff can, however, provide some information to help in rationalizing decision making. For example, if only limited funds are available and soil erosion is serious, it will be important to emphasize that little can be done to solve the problem unless a strong policy/support system (i.e., involving substantial resources) is potentially available. In other words, difficult decisions need to be made concerning the appropriate balance between need (i.e., equity) and potential pay off (i.e., efficiency) and the resources (i.e., usually governmental) available to address the problem. However, an important additional consideration in selecting a target area will be the degree to which a group/community are already, through their own efforts, showing interest and concern in tackling/solving problems related to soil conservation/land husbandry.

Start-up activities need to be initiated once the general location for a project has been selected. These can involve a number of actions, such as assembling and studying any information available on the area including any relating to land suitability and capability, obtaining the approval of local leadership to initiate activities, working out cooperative relationships and linkages with other agencies in and outside the area, assembling and, if necessary, training the staff (i.e., professional and support) to be involved in the activities, and making the necessary logistical arrangements.

4.2.2 Definition of Recommendation Domains

To be able to understand the context under which soil erosion occurs and to design feasible and practical strategies to ensure sustainability and control soil erosion, it is necessary to have a good understanding of the current farm-household system from the viewpoint of the land user. However, there is considerable heterogeneity in the farming population. This results from qualitative and quantitative differences in the biophysical and socioeconomic environment within which the farming population operates. However it is clearly impractical to address the needs of each individual farm household by means of separate FSD programmes. Hence it is necessary that farm households should be grouped together, and classified into recommendation domains. As indicated in Section 2.3, these are defined as homogeneous groups of farmers, who usually practice similar farming systems and face the same constraints and, most importantly, for all of whom the same solution(s) will be relevant.[46]

The constraints or problems defined in drawing up the recommendation domains will be dependent on whether a preventative or corrective soil conservation thrust is being contemplated:[47]

- In the ***preventative*** case, the 'felt' needs/problems of farmers will dominate and most likely will involve trying to identify solutions that will

[46] Of course, it is possible for all farming households to have the same major constraint (e.g., water erosion). However, the solutions to the problem are likely to be dependent on the resources they have available, which, in turn, influence the farming systems they practice. Therefore, because of differentiation with respect to potential solutions, a number of recommendation domains might be appropriate.

[47] Care needs to be taken in interpreting this statement. To some extent, this implies assumptions being made about what the problems seem to be. Farmers' perceptions -- both of problems and possible solutions -- may be quite different from the perceptions of the technicians or technologists. In general, if adoption of improved strategies is desired, and obviously it is, then farmers' views must prevail. This implies that perceptions of what should be done may need to be adjusted later, perhaps necessitating some modification in the definition of the recommendation domains.

increase short-term benefits to farmers. In its most benign form, adding the soil conservation dimension will simply involve adding sustainability criteria to the other criteria being used in evaluating the proposed solution(s).[48] Apart from this small adjustment, the approach and techniques used are likely to be almost identical to those commonly used in FSD work. On the other hand, actively seeking to enhance sustainability through conservation effective production technologies might necessitate some modifications in the recommendation domain definitions. For example, farming households that keep livestock as well as grow crops will have greater opportunities for designing strategies to increase interaction than those that have only crops. Therefore, although on every other basis both types of households could be in the same recommendation domain, the route to improving sustainability might well be different, indicating that the two farming households should be placed in different recommendation domains.[49]

- In the *corrective* case, the recommendation domains will not be determined so much by the 'felt' needs of farmers, although those needs and perceptions will be critically important in determining what might be adoptable. However, farming households in a particular recommendation domain will be those whose farms not only have the same type and degree of land degradation, which developed for similar reasons, but also

---

[48] It is regrettable that, to date, this has rarely been done.

[49] In the preventative case as indicated in Table 3, there is no description of, or assessment of causes, of soil erosion. However there is value in finding out from farmers whether they believe any of their land management practices are undertaken, primarily or partly, as a soil conservation measure. Likewise many of the farming practices they currently follow, for production and subsistence purposes, may be beneficial from a conservation point of view. Later (i.e., ***ex ante*** evaluation of proposed solutions and assessment of need for policy adjustments) it is important to ensure that the proposed changes will not disrupt these practices.

require the same land conservation strategy for their correction. Differences in the ability of individual households to sustain recommended corrective measure from within their own resources[50] will mean that they may have to be considered as separate recommendation domains and different corrective strategies developed for each.

The information required to classify farming households into recommendation domains derives mainly from informal unstructured[51] surveys using rapid rural appraisal (RRA) [Khon Kaen University, 1987] and participatory rural appraisal (PRA) [Chambers, 1992] techniques. These techniques are designed to be a comprehensive, but also quick and cost effective process to understand the determinants, operation, constraints, flexibility, and potential of the farm-household systems within their biophysical and socioeconomic environment. This exploratory descriptive/diagnostic work often involves interdisciplinary teams (e.g., a technical and social scientist) interviewing:

- Key informants, such as local leaders and others with influence within the community as well as knowledgeable individuals representative of various interest groups (e.g., village and government authorities, businessmen and religious leaders, women as well as men, and 'progressive' farmers).

- Groups of farmers who may not necessarily be statistically representative of the whole farming population.

---

50 For instance, farm households with limited cash and labour resources may find it difficult to maintain physical earthworks when the financial incentive that encouraged them to build them in the first place ceases on the termination of a donor funded project.

51 We include under this, so-called semi-structured surveys, in which a tentative list of topics or questions concerning which information is to be elicited, is prepared before hand.

- Heads and members of individual farming households.

Following the interviews, the data collected are discussed, analyzed, and interpreted in an interdisciplinary format. Finally, the results are fully documented and used as inputs into prioritizing the constraints (i.e., problems) and deciding on the potential solutions. These are major activities in the design stage.

Frequently, such informal work provides sufficient information and understanding to proceed directly to the design and testing stages. Undue delays in moving to these stages postpone opportunities for learning more about the farming environment as a result of trials [Raintree, 1987].

Sometimes however, collection of more concrete quantitative data may be necessary before moving on to design work. This type of exercise is designed to verify impressions gained in the more qualitative, exploratory descriptive/diagnostic work. Verification work traditionally has taken the form of formal structured surveys, suitable for statistical analysis. However, in the case of particularly complicated technical issues, this may need to be supplemented with other more direct techniques, such as topographic transects, soil and vegetation surveys, laboratory analysis of representative soil samples, monitoring/diagnostic technical studies, and trials to delineate constraints. This is particularly likely to be the case where corrective type soil conservation strategies are contemplated. As has been earlier emphasized, in such situations it is particularly important to rigorously assess the current causes and effects, and future risks, of soil degradation in order to develop soil conservation strategies that will prevent the problem being exacerbated or from recurring in the future.

Such transects, studies, and trials actually provide possible means, along with the more informal approach, for eliciting the information required in the next two steps. Also the information arising out of the next two steps (i.e., see

Table 3) can be very important in modifying the recommendation domains that have been tentatively identified, largely as result of the exploratory descriptive/diagnostic work. Consequently, as far as the corrective thrust is concerned, the distinction between the last four components in the descriptive/diagnostic stage can become blurred. With reference to the components as presented, they are designed to provide an understanding of the context within which the degradation processes are causing problems. However, where substantial soil erosion has already occurred, there may be merit in changing the sequence of addressing the different components so that the description of the farming systems and the definition of the recommendation domains follows the description of the degradation process. For example, the degradation process may be the same for all farms but the causes, and therefore, the potential solutions may be different, necessitating the creation of different recommendation domains.

Returning to the issue of verification, the general rule is that it should only be undertaken before the design stage when results arising from it are critically important in designing and evaluating (i.e., ***ex ante***) possible solutions. Otherwise, to prevent delays in moving through the various stages, verification work can be undertaken concurrently with work in the other stages. Such data are still useful in verifying the insights obtained during the exploratory descriptive/diagnosis exercise. Also such data can provide a useful quantitative database for farm and development planning, policy analysis, monitoring, and evaluation.

One final point with reference to collecting verification data is the issue of providing timely results. The efficiency of the link between data collection and analysis particularly needs attention, so that the data analysis can be completed shortly after the end of data collection. Microcomputer analysis permits rapid regrouping of the data according to major/multiple differences (e.g., recommendation domain, gender of household head, common constraints, etc.).

4.2.3 Description of The Degradation Processes

The need for a soil conservation programme is often recognized only after problems attributable to soil degradation have arisen. These problems do not simply involve degradation (i.e., such as soil loss or decreasing soil fertility) but also result in economic changes (e.g., such as loss of infrastructure, reduced agricultural productivity, and loss of investments and property). The need for a soil conservation programme will be determined largely as a result of these economic changes. An important task in the new approach to soil conservation is to precisely determine what has caused the degradation problems and why these have taken place or played a part in the process. This information is required so that appropriate soil conservation interventions can be planned and implemented to solve the problems. This and the next step are designed to help in that exercise.

A number of methods exist for obtaining the information required, including those mentioned in Section 4.2.2, experiences in other areas, and the conventional methods used by soil scientists. What is important to appreciate is that the causes may be determined by not only agroecological factors but also by those relating to the socioeconomic sphere. Consequently interaction with the farmer is critically important, not only in trying to elicit socioeconomic reasons but also in helping to try and relate them to those of an agroecological nature. In this endeavour farmers' evaluation and classification of land, soils, etc., may be particularly useful. What needs to be borne in mind in selecting the method or combination of methods is the objective of obtaining the data required in the least cost manner (i.e., in terms of both resources and time) commensurate with the degree of understanding and accuracy required to proceed to the design stage. The goal for having accurate data should not be an end in itself. A positive correlation generally exists between the degree of accuracy and the resources required to estimate a variable accurately. One of the most difficult psychological barriers to overcome in applying the FSD approach is to accept that relatively low levels of accuracy may be sufficient to provide the insights required to move to the design stage. This applies equally to soil conservation where qualitative rather than quantitative data may be all that is needed to determine soil erosion trends, it not being necessary to have precise figures on actual soil loss before designing technical solutions that fit the preventative thrust approach.

Therefore, the first step in this process is to describe the type, extent, and severity of the degradation processes that are playing a role in creating the economic changes, and vice versa.

4.2.4 Assessment of Causes, Effects and Risks of Degradation

A number of factors can play, or could play, influential roles in determining soil productivity. Depending on the circumstances, the role could be a positive one of enhancing rather than a negative one of decreasing soil productivity or causing the soil degradation process. Some of the factors are:

- Soil, climate, and topography characteristics.
- Land use and land management practices.
- Land, labour, and capital availability.
- Household needs and priorities.
- Farming strategies.

All of these need analysing, and the relationships between them should be assessed. This process should clarify which land users (i.e., farming families) are causing, or suffering from, the land degradation process. Some of the factors contributing to the land degradation problem are likely to be under the control of the farmers themselves, whereas others are not.

As a result of these considerations, the recommendation domains for corrective type soil conservation programmes should be more closely defined.

Now that a better understanding has been obtained about the causes of the degradation processes and their effects on the farming system(s) of the target group(s), it becomes possible to express the problem in terms of detrimental changes on conditions of soil and/or water, concentration of water, decline in soil structure, decrease in soil moisture availability, loss of soil fertility, runoff of water, loss of soil, etc. These are the problems that the corrective soil conservation strategies will need to address.

4.2.5 Ranking of Constraints (Problems) to be Addressed

Once the constraints and flexibility in the farming systems practised by farming households in each recommendation domain are identified, they are listed and ranked according to their restricting impact on production and the degree to which they inhibit farm-households fulfilling their aspirations. As indicated in the discussion of the descriptive/diagnostic stage, the character of the constraints delineated will be heavily influenced by what type of conservation thrust is being contemplated (i.e., preventative or corrective). Obviously, however, as we have emphasised earlier, all other things being equal, we believe emphasis should, whenever possible, be placed on improving land husbandry practices so important in implementing the preventative thrust.

4.3 DESIGN

4.3.1 Preliminary Selection of Potentially Suitable Solutions

For each constraint, feasible means are sought to alleviate, remove, or avoid it by exploiting flexibility in the farming system. Potential solutions that contribute most to fulfilling farm-household objectives, alleviate other constraints, are not costly to apply, are feasible to adopt, and interfere least with existing cultural values, are most likely to be acceptable to farming households.

When the constraints for which solutions are being sought relate directly to needs articulated by farming households, as is the case when the preventative thrust is being used, the solutions are more likely to be attractive. What is important is to ensure that the solutions proposed reap the requisite short-term benefits, while at the same time maintain or enhance the long-run productivity of the land (i.e., they are productive and conservation-effective). However, in situations for which the corrective thrust is to be implemented, the constraints do not necessarily reflect those articulated by the farming households. Therefore, the proposed solutions probably will not be as attractive. In such situations, as has been repeatedly emphasized, it is important to try to build in some short-run benefits for the farming households. This may be difficult to achieve where land degradation problems have become severe, and, as a result, strategies relating to the policy thrust may need to be applied (see Section 4.3.4). The degree to which such 'incentives' will be necessary will depend on the relative urgency of the degradation problem for the land users in relation to the other problems that they face.[52]

A broad range of possible solutions probably will be derived from the following sources:

---

[52] It should be borne in mind that where very severe land degradation has already occurred, although it may be technically possible to tackle the soil erosion problem through adopting the corrective thrust approach, it may not be cost effective to try and rehabilitate such areas. An underlying principle of the new approach to soil conservation is that the returns, to both farmers and the larger society, will be far greater from using scarce manpower and financial resources to implement soil conservation programmes that follow the preventative thrust approach. Examples of such programmes would be those that seek to sustain farm production in areas at risk of erosion, which have not yet been subjected to soil productivity decline.

- Local experience, for example, from farmers themselves, with their knowledge of useful indigenous techniques, etc.

- Experience elsewhere, for example, from other soil conservation programmes, from research stations, etc.

- Knowledge gained from 'initial testing' of prototype solutions (e.g., new forms of physical structure and/or vegetative cross slope barriers to combat soil erosion) or implementation of design type trials (e.g., to ascertain possible relevant tillage practices, or screen tree varieties for suitability, etc.).

In terms of the sources, the last one listed above should be avoided whenever possible. One reason is that it delays the move to the testing/implementation stage. Also, such work frequently can be undertaken more efficiently by other groups (e.g., researchers on experiment stations). However, these efforts can be justified when there are some unique features about the location that cannot be found easily elsewhere and are potentially influential in determining what the solution will be (e.g., particular soil type, specific weed complex, or a group of people with unusual social characteristics). Such work is usually managed and implemented by the researchers/ technicians themselves (see Section 5.3). However, although not directly involved in this work, groups of farmers can still contribute their perspective by commenting on the treatments and evaluating the results [Heinrich, Siebert et al, 1990]. We strongly believe in encouraging as much farmer involvement as possible, even at this stage, if the answers derived by researchers are to be anything better than 'hit or miss'.

Interdisciplinary groups consisting of team members and other knowledgeable people, including farmers, use their own wisdom plus information from the above sources in proposing solutions.

4.3.2 Ex Ante Evaluation of Proposed Strategies

This step involves evaluating the suitability of each potential solution in terms of the various constraints that were identified earlier. The proposed solutions, as emphasized in Section 3.7, have to be potentially: technically feasible; economically viable (i.e., in terms of both level and dependability of profit); and, socially acceptable. Although these considerations help determine the ability to adopt, a large number of other factors are likely to determine whether farmers will actually adopt the proposed technologies. Consequently, the validity and usefulness of this evaluation exercise depends on the involvement of potential adopters (i.e., farming households) and other local influential and knowledgeable people.

An assessment should be made of the impact of the proposed potential solutions not only on the farm-household system, but also in terms of factors important to society (e.g., equitability and sustainability).

The *ex ante* evaluation is undertaken in both biophysical and economic terms. For the economic analysis, a number of analytical methods, such as gross margin analysis, partial and whole farm budgeting, risk analysis,[53] and cost/benefit analysis, can be used. However, part of the assessment process involves considerations that go beyond these analytical techniques and include such factors as potential compatibility with farm-household goals, ease of adoption, managerial ability of farmers, gender impact, effects on employment, and social and cultural impact. Having farmers involved can be very important in helping to weight the relative significance of the different factors, especially when not all indicate potentially favourable results. Farmers will be in the best position to assess whether the anticipated

---

53 One of the simplest way of incorporating risk considerations is to do sensitivity analysis with respect to the partial budgeting exercise.

short-term benefits will be sufficiently attractive to enable implementation of soil conservation strategies with a long-run pay-off.

Indeed, it is the ***ex ante*** evaluation of criteria important to society (i.e., equitability and sustainability) that present special challenges. In the case of equitability, it is easy to develop technologies that can be adopted only by farming households with abundant resources. However, it is impossible to develop technologies that can be adopted only by resource poor farming households. In the case of sustainability, emphasis is obviously increasing on developing technologies that will maintain or enhance sustainability and physical structures that will reestablish it after degradation has taken place. Reasoning based on similar experiences obtained earlier, and in other locations, should provide some insights as to what might happen. However, a very important influence in determining the impact of the proposed solutions on equitability and sustainability will be the policy/support system. This is considered in the next section.

4.3.3 Assessment of the Need for Policy Adjustments

At this point, it is important to evaluate the proposed solutions in terms of answering the following questions:

- Are the current policies likely to positively or negatively facilitate the adoption of the proposed solutions, particularly with respect to the sustainability/conservation components?

- If policy changes are required to facilitate adoption, what are they?[54]

---

[54] Another question, not listed, but also relevant is in that assessment process to evaluate whether the proposed policy changes will discourage farmers from continuing conservation strategies they are already implementing.

- What are the implications of the proposed changes for government/society in terms of advantages (i.e., benefits) and disadvantages (i.e., costs)?

- Can government be influenced to change policies along the lines recommended?

If the current policies will facilitate adoption of a proposed solution, then all other things being equal, recommending that the solution(s) be further investigated, obviously has merit. However, if policy changes are required to facilitate adoption, then the other questions need to be addressed before proceeding much further. The proposed policy changes need to be evaluated very carefully in terms of whether or not it is reasonable for government to make the commitments required. For example, financial incentives for implementing conservation strategies may not be very attractive relative to cross-compliance type programmes designed to encourage both production and conservation. As a result, pragmatic compromises may be necessary. For example, because of costs involved, it may be more realistic to accept conservation strategies that slow down the soil degradation process rather than stop it entirely [Hudson, 1987].

A particularly difficult situation arises if government is unwilling or unable to implement the necessary policy changes. Under certain circumstances, there may still be merit in selecting the proposed solution for testing or implementing on a pilot basis and thus simulating the proposed policy changes. The circumstances under which this should be considered include the following:

- The proposed solution is particularly promising in terms of its impact on sustainability/soil conservation.

- A possibility exists that government may change its mind or aggressively seek external funding on the basis of demonstrated results.
- Resources are available to simulate the proposed policy changes on a pilot basis.

4.3.4 Final Selection of Strategies for Testing/Implementation

Finally, all the interested parties, including farmer representatives, need to reach agreement on which of the proposed solutions should be taken to the next stage for testing or implementation. Once again, because of the many factors that need to be considered in making the decision, farmer representatives must play a very important role. In view of the importance of policies facilitating or hindering the adoption process, there is also merit in trying to get inputs from planners/policy makers at this final step in the design stage, particularly if changes are likely to be necessary to encourage adoption.

At this point, a good ***ex ante*** understanding should exist of the potential contribution of the solution(s) accepted, in terms of solving the constraints articulated and contributing to sustainability and soil conservation.

4.4 TESTING/IMPLEMENTATION

4.4.1 Introduction

Initially decisions will need to be made as to whether the proposed solutions will be tested by farmers before implementation (i.e., dissemination) or whether the next step should be implementation. In general, the decision should be based on the following considerations:

- Solutions incorporating preventative soil conservation strategies should generally go through the testing phase. Exceptions would be made only in cases in which the proposed solutions have been adopted readily by farming households in the same recommendation domain in other areas, or substantial information already exists on their performance under practical farming conditions.

- Corrective type conservation strategies generally should be implemented immediately. There are two reasons for this:

    - A substantial amount of information is already available on physical structures for combating soil erosion once it has occurred.[55]

    - Because adoption of many of the strategies involves collective action on the part of a number of farming households, these strategies are not amenable to testing in the conventional sense.[56]

  If some doubt exists about the proposal, for example, whether collective action will work effectively, then a small pilot programme can be carried out first to assess the potential and possible implementation problems.

As a general rule, the higher the 'costs' (i.e., in terms of effort, money, etc.) and the longer the time-period before the stream of benefits starts, the more

---

[55] Some particularly useful references on these are Gallacher [1987], Hudson [1987], Moldenhauer and Hudson [1988], and Shaxson et al [1989], Sheng [1989].

[56] For example, testing activities usually involve individual farmers, replication across farms, etc., and are amenable to statistical testing.

important incentives of various types are likely to be in getting farmers to adopt the proposed solutions.[57]

4.4.2 Testing

On-farm testing provides the best forum for checking whether the proposed solutions will work. Initial testing work at the farm level may involve some managerial input from the 'outsiders' or technicians, although implementation is usually done by the farming households themselves. Later, if the technologies look promising, and as the farming households become more familiar with them, farmers themselves can assume responsibility not only for implementation but also for management.[58] Obviously, the last type of testing provides the best test of the technology, because the farming households in effect simulate incorporation of the technology into their own systems.

Testing is usually undertaken for a limited period (e.g., usually not more than three years). A favourable evaluation will mean that the technology is passed on for dissemination. An unfavourable evaluation will mean either adjustments being made in the technology, which is then retested, or complete rejection. The analytical techniques used are the same as those

---

57 Whenever potential solutions appear to involve high short term costs, with at best long term benefits, before considering the payment of any direct incentives, it is essential to be sure tht there are no alternative options that could be recommended which farmers might perceive as offering tangible benefits. Rather than promoting the soil conservation techniques contained in a conventional soil conservation manual (typically physical structures) it may be more appropriate to look for more innovative technologies and land use practices (eg. multi-purpose vegetative barriers and improved agronomic practices).

58 Differences in the types of trials are given in Section 5.2.

used in the *ex ante* evaluation process. However, much of the data used can be actual data collected in the testing process rather than the hypothetical data used in the earlier exercise. This is certainly true with respect to costs incurred and the benefits received during the actual testing period. Thus, the short-run net benefits can be evaluated easily. Because of the short time frame usually used for the testing exercise, the long-term benefits resulting from maintaining or enhancing sustainability are not taken into account in the evaluation process.[59]

However, because many of the costs contributing to these long-term benefits are likely committed during the actual field level testing period, the short-run net benefit period, in fact, will be underestimated. If the technology is still attractive to farming households, in spite of this, there should be no problem about adoption, with sustainability incorporated in the process.[60]

---

59 Whereas the usual testing period may be too short to determine whether the technology is maintaining or enhancing long term sustainability, there is considerable scope for the use of simple visual parameters to obtain a qualitative estimate of its conservation effectiveness. Thus simple observations repeated at regular intervals during the testing period (eg. every year or at 2-3 weeks intervals during a cropping season) can be used to assess the underlying trend and to determine whether the technology has the potential to prevent, or reduce, soil degradation. For instance, changes in topsoil colour and structure can be used as indicators of changes in soil organic matter levels. Regular observation of the surface condition of a farmer's field for signs of crusting and rill development will indicate whether rainfall impact and surface runoff are still causing soil ersosion. Measuring the height difference between the uphill and downhill sides of a conservation structure, contour hegerow or grass strip will reveal whether soil is accumulating on the uphill side and therefore indicate its effectiveness in trapping soil withing the plot.

60 If costs actually continue at significant levels into the future, benefit/cost analysis is really the only valid tool for evaluating the technology. The problem with this is that future costs and especially benefits are not directly measurable, and farmer households

These types of quantitative analyses with hard objective data need to be complemented by softer, more subjective, qualitative, attitudinal data from the farmers participating in the testing process. Such information can be very helpful in ascertaining whether farmers are likely to adopt the proposed strategies in the absence of policy changes. If policy changes are necessary, such as reducing fertilizer subsidies to discourage excessive use or implementation of a cross-compliance policy, then the technology should not be extended until required conditions are met. Also, before forwarding the proposed recommendation for formal approval by the necessary authorities prior to passing it to extension for dissemination, it is important to include some extra information. The object is to help as many farmers as possible successfully adopt the recommendation. The following types of information are particularly useful:

- ***Targeting information*** that states under what environmental (i.e., both biophysical and socioeconomic) conditions the recommendation will be most relevant (e.g., certain cover crops might be most suitable on specific soil types and for farms using forage for livestock).

- ***Conditional clauses*** that state what to do if it is not possible to follow the recommendation exactly as instructed (e.g., stepwise approach to the adoption of a recommended package, don't apply top dressing of fertilizer if the first weeding was not done satisfactorily, don't construct

living at the subsistence level will discount the future very heavily. Therefore, even if a high level of internal rate of return is estimated, such households are unlikely to be interested. At the same time, if there is reasonable confidence concerning the estimate, this could help encourage government to implement, where appropriate, some incentives to encourage adoption.

graded earthbanks to control surface runoff if this cannot be discharged into stable waterways, etc.).[61]

4.4.3 Implementation (i.e., Dissemination)

Some differences are likely to occur in the approaches used with respect to conservation strategies relating to the types of thrusts. These include:

- Technologies that relate to the preventative thrust, which have successfully passed through the testing step, will usually be disseminated through the extension service. Such technologies can be adopted on an individual household basis and will be directed to farming households in specific recommendation domains. At this point, there is little, further, day-to-day involvement of staff responsible for developing the technologies.

- Physical structures that relate to the corrective thrust will tend to be implemented by specialist soil conservation staff, in conjunction with extension staff and the farmers involved. Although some physical structure type strategies may involve farmers on an individual household basis, many probably will involve collective agreement. Indeed, this is preferable if collective action is involved, so that a feeling of collective responsibility is engendered and maintained. Unfortunately, FSD practitioners have generally had little experience in such initiatives. Therefore, ideas on approaches to securing and maintaining cooperation

61 This implies that research has a huge job to do in modifying recommendations for non-optimal situations. However, we believe much of this information can be provided on the basis of common sense derived from experience rather than from exhaustive experimentation.

among farming households need to be obtained from other sources (e.g., community development, recent experience in agroforestry work, etc.). As indicated earlier, a small pilot project may be desirable initially, if full-scale implementation will require policy changes on the part of government (see Section 4.3.3). Such an approach also would be desirable initially, for example:

- If some testing of methods for obtaining and maintaining collective action and responsibility is desirable.
- If some question exists as to whether lack of land tenure security may inhibit land user cooperation.
- Whether land user labour will be available to participate in construction activities at slack periods in the agricultural cycle, etc.

Full-scale implementation can take place once these issues are satisfactorily settled.

4.5 IMPLEMENTATION/DISSEMINATION IMPACT

4.5.1 Monitoring Degree of Adoption

This final stage provides the only true ***ex post*** evaluation of the strategies designed to ensure increases in short-term benefits, while at the same time at least maintaining, or preferably improving, sustainability and/or soil productivity.

Knowing whether research efforts have been fruitful (e.g., for justifying continued funding) is always valuable. However, other important reasons exist to monitor what has happened with respect to the implementation/dissemination of the strategies. Two of them are as follows:

- Hard quantitative information on the soil conservation benefits of the strategies will become available only after a period of years. Obtaining such information enables estimates of the benefit/cost ratios to be more accurate. Such information can be very useful in future design-type activities.

- Monitoring gives an indication as to whether implementation/ dissemination has been successful or whether some changes are required. For example, in the case of:

  - Technologies involving prevention type soil conservation strategies -- what has been the rate of adoption, have modifications been made when adopted by farming households, has the level of expected returns been maintained, has land productivity been maintained/enhanced, etc?

  - Physical structures involving corrective type soil conservation strategies -- have they survived over time; where necessary, are they maintained on a regular basis (e.g., terraces, bunds, etc.); if collective action and responsibility are required, are they still functioning, has soil erosion been stopped, is soil productivity increasing, etc?

Favourable answers to the above questions can provide greater confidence in advocating similar solutions to analogous constraints in future design type activities. If some of the responses are not favourable, then changes may be necessary, preferably after determining a proper understanding of the reasons -- usually through in-depth dialogue with the farmers. Possible changes are briefly outlined in the next section.

4.5.2 Proposed Modifications

Information arising from the monitoring process can be particularly useful, even after a number of years, in determining whether the strategies earlier disseminated should:

- Continue to be implemented/disseminated.
- Should continue to be implemented/disseminated only after further modification or changes in the policy/support system or the technology itself.
- Should no longer be disseminated.

5. IMPLEMENTING THE NEW SOIL CONSERVATION APPROACH: STEPS, TOOLS AND TECHNIQUES

5.1 INTRODUCTION

Implementation of the new soil conservation approach utilizing the principles of FSD, will require major changes on the part of those used to the more conventional approaches to soil conservation. Three particularly important ones are as follows:

- It is to accept that the farmer is in the 'driver's seat' as far as any changes are concerned. Thus, he/she has to be treated as a person who can interact constructively and cooperatively with outside experts. He/she can no longer be treated as an object, or as part of the problem, but as a person who can suggest and implement strategies that can help solve the problem. For this to occur, the soil conservation specialist has to become an expert in communication, and work to develop his/her credibility in the eyes of farmers.

- The new approach, which is more 'holistic' in nature, requires thinking in more of a systems or interdisciplinary context and necessitates the collection and analysis of many different types of data (i.e., both technical and socioeconomic in nature).

- Because of the need for greater amounts and types of data, much of which are obtained directly from farmers, unfamiliar methods are required for collecting and analyzing the data. In particular, in the interests of using research resources in a time-efficient manner, much greater reliance needs to be placed on informal survey (RRA/PRA) methods, rather than on direct measurement or an observation format for collecting data. At the same time, whenever possible, cheaper qualitative/ordinal/subjective rather than quantitative/cardinal/objective types of data need to be utilized.

The objective of this chapter is to provide a brief introduction to some of the steps, tools and techniques that can be used in implementing the new approach to soil conservation using many of the principles of FSD. The material in the chapter should help sensitize the soil conservation specialist to some of the techniques that are likely to be unfamiliar to him/her. At the same time, none are discussed in detail, because these plus others are discussed in a large variety of methodological manuals on farming system type activities.[62] Also, because this publication endorses the approach used in FSD, the manuals written with respect to this should be of particular value [FAO, 1989 and 1990B]. In addition there are a number of papers that provide particularly useful methodological information that is closely related to FSD and various aspects of soil conservation.[63]

Before looking at these techniques in a little more detail, we express the components of the conceptual framework outlined in the last chapter, in more specific steps that we suggest are followed in order to provide a systematic

62 Directly related to much of the material presented in this chapter is a paper produced by Douglas [1992], which is mentioned in the foreword to this publication. Other useful references that apply more directly to FSD itself are Amir and Knipscheer [1989], Andrew and Hildebrand [1982], Caldwell and Walecka [1987], Caldwell, Taylor et al [1987], Caldwell, Frankenberger et al [1987], CIMMYT [1988], Frankenberger and Walecka [1987], Hildebrand and Poey [1985], Norman and Collinson [1986], Shaner et al [1982], Tripp and Woolley [1989], Worman et al [1990], Zandstra et al [1981].

63 Particularly useful ones are by Douglas [1990] and a number produced on agroforestry by the International Council for Research in Agroforestry (ICRAF) [Raintree, 1987; Rocheleau, 1985 and 1990]; Muller and Scherr [1990].

and efficient approach to addressing the thrusts we have articulated in the new approach to soil conservation.

5.2 SPECIFIC STEPS IN ADDRESSING THE NEW APPROACH TO SOIL CONSERVATION

A list of steps that we suggest can be used in applying the approach to soil conservation is given in Table 4. It is a slightly modified version of an even more detailed version given in Douglas [1992]. We would like to stress three points about the list:

- It is critically important that the farmers themselves are involved whenever possible -- indeed their involvement can be very helpful in the quick and efficient movement between the steps.

- It is indicative of what needs to be done but does not necessarily always have to be followed to the letter -- modifications may be desirable depending on the circumstances.

- Some of the steps can perhaps be skipped if the corrective rather than the preventative thrust are being addressed -- with reference to the discussion on this see Table 3 and Section 4.1.

In the following discussion we have tried to be brief. Some of the points are self-explanatory, and need little further discussion. However, those wishing to have any more information about any of the topics should consult Douglas [1992].

Also before getting into a discussion on specific points and issues, we preface our remarks by devoting the next section to a brief but more generalized discussion on data collection methods.

TABLE 4: STEPS NECESSARY IN APPLYING THE NEW SOIL CONSERVATION APPROACH

Phase	Stage	Activities

A. Description/Diagnosis

A1 Planning and Preliminary Investigation

A1.1 Preparatory planning
- o Ascertain national and local land use planning goals/objectives
- o Decide location, extent and boundaries of work
- o Identify concerned agencies and officials, plus potential collaborating institutions and staff
- o Identify necessary disciplines and interdisciplinary planning team formation
- o Determine the time, resources, and methods available for exercise
- o Establish contacts with, and brief, community leaders, politicians and government officials in study area

A1.2 Collection and review of secondary data
- o Identify, collect and review available secondary data
- o Identify major information gaps

A1.3 Field reconnaissance
- o Reconnaissance familiarization field visit by team to area
- o Initial consultation with local 'key' informants.

A1.4 Preliminary characterisation of farm household circumstances
- o Identify key natural environmental circumstances affecting land use suitability

TABLE 4 (Continued):

Phase	Stage	Activities
	A1.4	(Continued)
	o	Differentiate and describe land management units with dissimilar ecological conditions requiring separate consideration
	o	Identify dominant land use enterprises in area
	o	Determine key socioeconomic circumstances of farm households
	o	Tentatively identify homogeneous groups of farm households (recommendation domains)
	o	Tentatively characterise dominant recommendation domains according to resources available, land use enterprises, constraints, problems and potentials
	o	Identify information gaps and prepare terms of reference for the primary data collection stage
A2	Assessment of farm houschold circumstances and problem identification	
	A2.1	Primary data collection
	o	Informal survey
	o	Formal natural resources and socioeconomic surveys (if necessary)
	A2.2	Analysis and description
	o	Analysis of field collected data to determine natural and socioeconomic situation of farm households
	o	Identify key production and conservation (sustainability) problems, their causal factors and constraints
	o	Determine and provide description of each of the major recommendation domains
	o	Identify potential intervention points within the farming systems through ranking constraints and identifying related causal factors.

TABLE 4 (Continued):

Phase	Stage	Activities
	A2.3	Broad scale considerations for development of solutions
	o	Determine level of intervention -- on farm, within wider community or at national policy level
	o	Determine resource needs and availability -- within household, community, region or nation
	o	Determine development approach (e.g., adaptation of, substitution for, or addition to existing land use practices)
	o	Determine what any solutions are required to do
	o	Determine nature and type of limiting constraints -- are they fixed or surmountable?
B Design		
B1 Development of potential solutions and extension recommendations		
	B1.1	Review of land use enterprise/farm management options
	o	Appraise suitability of existing land use enterprises and farm management recommendations and determine reasons for adoption or non-adoption
	o	Identify and review possible alternative land use enterprise/farm management options
	B1.2	Selection of potential land use enterprises/farm management options
	o	Brainstorming potential options
	o	Produce list of potential technical and policy options for solving farmers' production and conservation (sustainability) problems
	o	Review and select most promising options for further consideration and detailed development

TABLE 4 (Continued):

Phase	Stage	Activities
	B1.3	Development of technical and policy action proposals
	o	At field level, develop detailed specifications for field level testing/implementation of most promising land use enterprise/farm management options including, as appropriate, possible alternatives, from which individual farm households can select those perceived as suited to their specific circumstances
	o	At policy level, consider changes required to current policy environment at local, regional and national level, and formulate and communicate specific policy recommendations
	B1.4	***Ex ante*** appraisal and possibly revision
	o	Have proposed solutions appraised by interested parties including farmers and extension agents
	o	Evaluate ***ex ante*** proposed solutions in terms of technical, socioeconomic and sustainability criteria
	o	If necessary revise and modify the proposed solutions and re-evaluate
C. Testing/Implementation		
C1.		Testing and/or implementation of solutions
	C1.1	Determine research needs
	o	Review current knowledge, attitudes and experience to determine if can immediately be implemented, or need to be further tested
	o	If further testing required determine what, where and how it should be done

TABLE 4 (Continued):

Phase	Stage	Activities
	C1.2	Research and extension planning
	o	Develop overall plan of action detailing research investigations and extension programmes, activities and messages
	o	Agree on linkages and interaction between farmers, researchers and extension
	o	If necessary determine financial and human resource requirements and the time frame for the activities and submit to the appropriate authorities
	C1.3	Implementation of research and extension activities
	o	Establishment and execution of trials
	o	Periodic participatory workshops/field discussions of trials and, if necessary, make necessary adjustments
	o	When ready, make necessary recommendations on technologies to extend together with any adjustments that are required in policy/support systems
	o	Dissemination of extension recommendations to farmers
D. Dissemination Impact		
D1.		Monitoring, evaluation and revision
	D1.1	Monitoring and evaluation
	o	Monitoring of progress with dissemination of the extension recommendations, and implementation of policy and enabling legislation recommendations
	o	Evaluate farmers' reasons for adoption, modification or rejection of the extension recommendations

TABLE 4 (Continued):

Phase	Stage	Activities
	D1.1	(Continued)
	o	When appropriate, evaluate the effect of any changes in the policy/support system environment
	D1.2	Revision of research and extension activities, and policy/support systems
	o	If necessary feedback further priorities to research and modify the extension recommendations
	o	If required push for adjustments of the policy/support system

5.3 GENERAL BACKGROUND ON DATA COLLECTION METHODS

As we have repeatedly stressed, both technical and socio-economic data are needed in the new approach to soil conservation. For example, the time spent on ploughing a field is dependent on the type of draft used and the number of hours per day that it is used. However, ploughing time is also determined by a number of technical factors, such as moisture in the soil and the soil type. Unfortunately, data collected and analysis undertaken often reflect only socio-economic or technical data. Looking at only one aspect does not represent the whole picture. Collecting both types of data is not always possible, but the limitations of collecting only one type should be recognized.

The most common methods of collecting data in FSD work that can be directly applied in the implementation of the new approach to soil conservation are the following:

- Surveys (interviews and questionnaires).
- Observation.
- Direct measurement.

Surveys, which involve interviewing, include many different types. Direct observation[64], because of its expense, has tended to be used sparingly in FSD work, and when utilized, has tended to be undertaken in conjunction with other methods. Direct measurement activities are often associated with monitoring and trial activities.

A number of factors influence what method or methods of data collection are used. The ability of different methods to address these is evaluated in Table 5. This evaluation, admittedly done in a somewhat subjective manner, involves ranking each factor or operational constraint as to how well they are dealt with by the different methods. A ranking of 1 to 6 is used, with '1' indicating a very favourable ranking and '6' indicating a very unfavourable ranking.

Because of different disciplinary needs, as well as the need to optimally use the limited research resources, a strategy consisting of a combination of methods is normally used. The results in Table 5 show that no one method of data collection is basically superior in minimizing operational constraints.

Careful thinking is required about how accurately the data must be collected. The more accurately it is collected, the more expensive it tends to be. The costs (e.g., financial and time) associated with a data collection method should be comparable to the type of information sought. For example, multiple-visit surveys for quantifying actual seasonal labour flows are

[64] In the sense of systematically observing and recording farm-household activities, however, see section 5.4.5 for details of its application in an informal survey context.

TABLE 5: EVALUATION OF DATA COLLECTION METHODS IN TERMS OF OPERATIONAL CONSTRAINTS[a]

CONSTRAINTS	STRATEGIES				
	INFORMAL SURVEY	FORMAL SURVEY ONE/FEW VISITS	FORMAL SURVEY MANY VISITS	OBSERV-ATION	DIRECT MEASURMENT
Finances	1	2	3	6	5
Staff skills	6	3	3	6	5
Time required	1	1	5	6	
Type of system:					
Simple	1	2	1	1	
Complex	5	5	2	1	1
Errors:					
Sampling	6[c]	1[b]	3[b]	5[c]	5[c]
Measurement: [d]					
SPRD[e]	3	3	1	1	3
CNRD[f]	6	6	2	1	1

a. Evaluation of data collection strategies in terms of cost per unit with respect to various operations (1 = lowest cost per unit, 6 = highest).

b. Evaluated in terms of ability to reduce sampling error.

c. Evaluated in terms of ability to specify sampling error.

d. In the following two lines:
'Single point' to 'continuous' relates to an activity over a brief time span compared with one that has a much longer time dimension.
The 'registered' to 'non-registered' refers to an event that leaves a record or vivid recollection compared with one that doesn't.

e. Single-point registered data.

f. Continuous non-registered data.

Source: Modified from Kearl [1976].

expensive and require a good deal of time. A rough estimate based on a much cheaper and quicker, one-shot survey in fact may be sufficient for the purpose required. It is important to consider whether a more accurate measurement would improve the understanding enough to justify giving up other opportunities, such as working with more farming families.

For surveys, measurement errors are often reduced with more frequent interviewing techniques, whereas sampling errors are reduced by involving larger numbers of farmers. Because, there are often limited resources for undertaking FSD and related activities, it is usually necessary to make a trade-off between these two errors.

Related to the above and to efficiency is the idea that sometimes cheaper, qualitative type data may provide sufficient understanding rather than having to rely on expensive, quantitative type data. Qualitative information can include not only attitudinal information, but information on the relative labour requirements of different operations, etc. Limiting quantification to key characteristics reduces the cost involved in collecting data.

Table 6 is an attempt to indicate the major data collection methods now considered by most FSD practitioners to be best during each phase of FSD and hence for the new approach to soil conservation. Once again there is a degree of personal subjectivity about the degree of importance attached to each method. Indeed some of the weighting reflects what may be considered to be desirable rather than what is always done. As examples, multiple visit formal surveys used to be very important in the descriptive/diagnostic stage while a one visit formal survey is still considered to be important by some practitioners in the same stage. Also impact/adoption activities are still not undertaken in the dissemination phase to the extent that would be desirable. It is therefore important to note that there have been changes in priorities given to different methods over the years. Reasons for this relate to evolution in methodology and increasing concern with maximizing the return

TABLE 6: CURRENTLY USED DATA COLLECTION METHODS BY FSD PHASE[a]

APPROACH	DESCRIPTIVE/ DIAGNOSTIC	DESIGN	TESTING	DISSEMINATION
Secondary information	1	1		2
Dialogue with knowledgeable people	1	1		2
Data collection method:				
Informal surveys	1	3	2	2
Formal surveys:				
One/few visits	2		3	1
Multiple visits				
Observation	3			3
Direct measurement	5	4	1	4

a. Those efforts not involving primary data collection are separated out in the table. Also within each phase, an subjective evaluation is given of the importance of each method with 1 indicating most important. A value of 5 indicates that it is generally undesirable and should only be undertaken in extreme situations. Values in between 1 and 5 reflect the gradient between important/desirable and unimportant/undesirable. Cells with no entry reflect little or no reliance on that method/approach.

from limited research resources. Related to the latter have been three considerations, namely:

- Not aiming for near perfect understanding and great accuracy but doing enough to get the job done.[65] As just discussed, attainment of perfect understanding and accuracy would involve considerable expense.

- Moving through the different stages of FSD as quickly and efficiently as possible. Obviously this helps in improving the return from limited research resources.

- Development of better but not necessarily the perfect solutions to problems. All farmers are different in some respect and the perfect solution, or near perfect solution, is an impossible goal idealistically and practically in terms of the limited developmental resources available.

Surveys involving interviews are used a great deal in FSD type activities, and as implied in Table 6, also should be very important in implementing the new approach to soil conservation. Figure 4 illustrates the different types of surveys (i.e., interviews) that are possible, whereas Table 7 compares the characteristics of the two main groups of surveys, unstructured (informal) and structured (formal). Structured interviews involve collecting information on the same topics from all respondents. In contrast, unstructured interviews involve collecting different information on different topics from different respondents -- individually or in groups.

---

[65] Chambers [1985] refers to this as optimal ignorance and appropriate imprecision. Optimal ignorance reflects the fact that, in order to formulate hypotheses in a limited period of time, it is not necessary to collect data on everything. Chambers describes it as the importance of knowing what facts are not worth knowing. Appropriate imprecision is based on the premise that, especially in formal surveys, much of the data collected and the analyses undertaken, have a degree of accuracy which is unnecessary. Orders of magnitude and direction of change are often all that is needed.

TABLE 7: COMPARING GENERAL CHARACTERISTICS OF FORMAL AND INFORMAL SURVEYS

CHARACTERISTIC	INFORMAL	FORMAL
Background information required	Less	More
Time allocation by researchers:		
Preparation	Less	More
Implementation	More	Less
Analysis/writing up	Less	More
Total time	Less	More
Hypotheses: Required beforehand	Not essential	Essential
Created during	Yes	No
Likely discipline interaction	More likely	Less likely
Implementation:		
Questionnaire used?	No	Yes
Interviewers	Researcher(s)	Mainly enumerators
Potential for creativity/iteration	Maximum	Minimal
Potential for learning/verification	Mainly learning	Mainly verification
Potential for representative sample	Less likely	More likely
Potential quality of information:		
Attitudinal	Better	Poorer
Qualitative	Better	Poorer
Quantitative	Poorer	Better
Probability of high:		
Sampling errors	Higher	Lower
Measurement errors	No difference	No difference
Value of statistical techniques in analysis	Little	Great

In structured interviews, the information to be collected is decided prior to the interview. However, in unstructured interviews, the types of information collected tend to be partly stimulated through interaction between the interviewer and respondent.

In FSD, a great deal of emphasis is placed on informal surveys (i.e., particularly using RRA and PRA techniques), supplemented by single-visit formal surveys, particularly during the descriptive/diagnostic and dissemination/impact stages. However, collection techniques involving multiple visits and direct measurement data are often used in conjunction with implementation of trials during the testing stage. For those sceptical about the value of informal surveys, points to bear in mind are:

- It is possible to have a very quick turnaround in terms of results that can be fed into the next stage.
- Ways are available to minimize the levels of inaccuracy and bias in the information that is collected -- see, for example, Muller and Scherr [1990].
- If measurement error is not a major issue (i.e., a general understanding rather than actual figures is important), then according to Table 5, single-visit surveys look very favourable in minimizing operational constraints.
- Many farming systems practitioners have found that the flexibility and dialogue features of informal surveys make them very useful techniques for obtaining information on all sorts of topics.[66]

---

[66] For example, Muller and Scherr [1990] have concluded that informal surveys are superior to other methods in obtaining information on many topics (i.e., biophysical description, farmer needs and problems, management practices, farmer attitudes, knowledge and interest, economic information, etc.).

FIGURE 4: TYPES OF SURVEYS (INTERVIEWS)

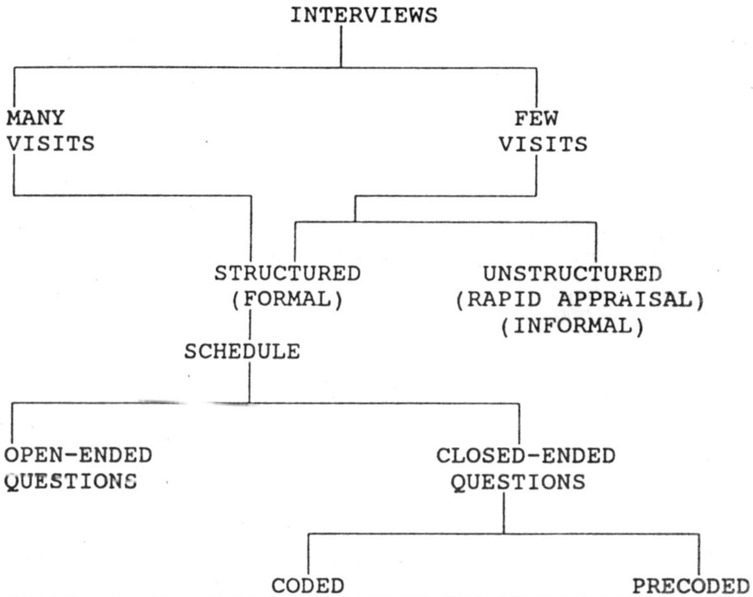

Source: Modified from Kearl [1976].

For all types of data collection, it is important to find ways to make the results available as quickly as possible. This is done through improving the efficiency of the data collection, processing, and analysis links. All surveys, no matter how they are undertaken, need to be designed to facilitate quick processing, so as to simplify the transfer of data to computer-based systems.

TABLE 8: EXPECTATIONS OF DIFFERENT TYPES OF TRIALS[a]

Item	Researcher Managed and Researcher Implemented (RMRI)	Researcher Managed and Farmer Implemented (RMFI)	Farmer Managed and Farmer Implemented (FMFI)
Experimental:			
Stage:	Design[b]	1st stage testing	2nd stage testing
Design:			
Complexity	Most	Less	Least
Type	Standard	Simple standard	With and without
Replication	Within and between sites	Usually only between sites but can also be within	Between sites only
Who selects technology?	Researcher	Researcher/farmer	Farmer
Participation by:			
Farmer	Least	More	Most
Researcher	Most	Less	Least
Numbers of farmers	None	Some	Most
Farmer groups	Least	More	Most
Potential:			
"Yield"	Most	Less	Least
Measurement errors	Least	Greater	Most
Degree of precision	Highest	Less	Least
Data:			
Hard (objective)	Most	Less	Least
Soft (subjective)	Least	More	Most
Determine cause/effect relationships	Easiest	Less easy	Least likely
Incorporation into farming system	Least	More	Most

TABLE 8 (Continued):

Item	Researcher Managed and Researcher Implemented (RMRI)	Researcher Managed and Farmer Implemented (RMFI)	Farmer Managed and Farmer Implemented (FMFI)
Evaluation:			
By whom?	Mainly researcher	Researcher/farmer	Mainly farmer
Nature of test	Assesses technical feasibility feasibility	Some of each plus economic evaluation	Validity for farmers-practicality, acceptable
Appeal to:			
Researchers	Most	Less	Least
Extension Staff	Usually least	More	Most
Farmers	Least	More	Most
Ease of acceptance of results of trial	Researcher	Researcher/farmer/ extension	Farmer

a. Some of the entries in the table have a certain degree of subjectivity, but generally they reflect the usual case. In a sense, these expectations also reflect the reasons why the different types of trials are undertaken.
b. Standard multi-locational trials are also RMRI.

Source: Norman [1989].

If structured surveys are implemented, schedules with pre-coded, closed-ended questions are potentially most efficient.[67] Such surveys should be used when:

67 Another factor relating to reducing the turn-around period is, whenever possible, implementing one-shot formal surveys that have limited objectives (i.e., special subject surveys). This is in contrast to multiple-visit formal surveys with many objectives,

- The researchers themselves are not conducting the interviews.
- The enumerators are not particularly well-qualified.
- Comparative analysis, such as cross tabulation, is to be undertaken.

Turning to direct measurement, trials play a very important role in the testing/implementation stage. To be effective in their work, FSD practitioners and for that matter, those involved in the new approach to soil conservation need to interact with farmers, station-based researchers, extension and development agency staff, and planners. Similar types of trials do not appeal equally to all these clients. As a result, three different types of trials are often used. These trials can be differentiated on the basis of who manages and who implements the trial, that is, researcher (i.e., technician) or farmer (Table 8):

- Researcher managed and researcher implemented (RMRI).
- Researcher managed and farmer implemented (RMFI).
- Farmer managed and farmer implemented (FMFI).

Another type of trial, called superimposed, also can be very useful. Superimposed trials include elements of management by both farmers and researchers, with implementation primarily the responsibility of farmers. Therefore, the trials generally involve superimposing a degree of research management on a plot or animal being managed and worked on by the farmer. These trials tend to be single-factor experiments, such as looking at the response to fertilizer superimposed directly on a farmer's own plot where he/she is providing both labour and management. Superimposed trials can be pre-planned or unplanned in the sense of responding to a problem that has arisen and for which a satisfactory solution is sought. This contrasts with the other types of trials mentioned earlier, which are implemented according to a plan drawn up earlier.

Table 8 also indicates that the three trial types can be differentiated in terms of research objectives, methods, experimental design, types of data collected, methods of analysis, and evaluation criteria. Often in FSD type work, and in the new approach to soil conservation, technology design begins at the RMRI level. Technologies thought to be appropriate are then passed on for testing with the farmer, first at the RMFI level and then at the FMFI level. Obviously, the farmer becomes progressively more involved as one moves from experimentation at the RMRI to the FMFI level. Farmer groups [Heinrich, Masikara et al, 1990] are particularly useful for implementing trials at the FMFI level. They also play a role at the RMFI level. In RMRI type trials, undertaken on-farm, the researcher input is much greater. However, even here, it is possible to elicit farmers' opinions by bringing a group together for the purpose of obtaining their opinions on possible treatments and the results that arise as a consequence of implementing the trial [Heinrich, Siebert et al, 1990].

RMRI trials on farmers' fields are analogous to those conducted on the experiment stations. Therefore, the level of testing achieved meets the standards demanded by experiment station-based researchers. However, FMFI trials are the most appealing to the farmer and provide the most practical test of the technology. Because of management and resource constraints, yields or returns will diminish from the RMRI to the FMFI level.

Information in Table 8 notes the major differences between RMRI work, mainly the preserve of experiment station researchers, and RMFI and FMFI trials that are emphasized in on-farm work. One such difference, for example, is that cause-effect relationships and 'hard' (i.e., quantitative) objective-type data are more easily obtained from RMRI work. In contrast, farmer attitudes and inputs into the research process are more readily obtained from RMFI and FMFI work undertaken on their own fields. However, data collected under farmer-implemented formats, particularly at the FMFI level, are likely to be 'softer' (i.e., more qualitative) and more subjective in nature. Moves towards greater incorporation of the farmer, the ultimate customer of FSD work, in the research/evaluation process, therefore, usually mean a relative increase in 'softer' qualitative type of data. Unfortunately, this type of data is less acceptable to many government-related individuals who are responsible for approving recommendations for official dissemination.

All three types of trials are likely to be vitally important in contributing to a well rounded programme involving the new approach to soil conservation. Once the purposes of the different types of trials are understood, complementarities that exist between them can be recognized easier. Adopting appropriate criteria for evaluating their worth also is easier. For example, it is unreasonable to expect the same low level of variation in the results of FMFI work as in those of RMRI work. This is because in FMFI work, there is much less control on the part of the researcher. Therefore, measurement errors are likely to be much higher and the degree of variability in the levels of the non-experimental variables is also likely to be greater.

What we have just discussed with respect to the different types of trial has a linear connotation -- that is, from researchers to farmers. However, this is no longer strictly accurate, since an evolution in thinking and implementation is currently taking place. For example:

- Farmers are increasingly being involved much earlier in the research process than used to be the case, and RMRI trials are only undertaken by FSD practitioners when absolutely necessary.

- Methods are increasingly being advocated that should enable the needs of more than one client (i.e., researchers and farmers) to be better satisfied in FMFI type trials. Farmers' needs are more likely to be satisfied as a result of their being involved as much as possible in such trials. At the same time researchers' needs can now be better met through alternative statistical techniques such as modified stability analysis [Stroup et al, 1992] and, expressing farmers' evaluation or assessment in a somewhat more quantitative manner through the use of PRA techniques.

We hope that practitioners of the approach to soil conservation will keep abreast of these developments and incorporate them as they become more widely used.

With this general introduction on data collection methods we now turn to a more focused discussion on the implementation of the different phases of the new approach to soil conservation, and suggestions on some of the steps, tools and techniques that can be used in accomplishing the tasks that need to be done.

5.4 THE DESCRIPTIVE/DIAGNOSTIC PHASE

We suggest in Table 4 that activities during the description/diagnostic phase can be divided into two major stages (i.e., A1 and A2) and seven different steps (i.e., A1.1 - A1.4, A2.1-A2.3). The actual activities that are required are fairly self-explanatory. A few specific points are covered in the following sub-sections.

5.4.1 Preparatory Planning

Obviously the new approach to soil conservation has to be involved in designing initiatives that are compatible with the goals/objectives of the national and local plans for the use of land. We earlier discussed that the areas where such soil conservation work will be implemented will also be heavily dictated by 'outside' factors or forces (see Section 4.2.1). At the same time there are compelling reasons for those responsible for implementing the new approach to soil conservation to have some role on deciding the geographical boundaries once the general target area has been selected. For example, in some types of soil conservation, if cooperation is required over a specific area of land or among a specific group of people, the relevant boundaries may be determined by a natural watershed[68] and/or a specific community of people.

The characteristics and needs of the target area dictate to a great extent those agencies and institutions that the groups responsible for implementing the new approach to soil conservation need to collaborate with. The reasons for such collaboration will not only be determined by what those agencies and institutions can do in helping solve the problems of the target area but also in helping overcome any deficiencies in the interdisciplinary team responsible on a day-by-day basis for implementing the new approach to soil conservation. Discussion on issues relating to this cooperation/collaboration are considered briefly in Chapter 6. What is important to emphasise is that the team functions not in a multidisciplinary mode but rather one that is interdisciplinary in approach. A multidisciplinary mode involves the active participation of a number of different disciplinary specialists, but each specialist tends to plan, execute and evaluate separately his/her component of the overall plan -- something that has been common, for example, in more conventional approaches to watershed management planning. In contrast, if the new approach to soil conservation is to be successfully implemented, the interdisciplinary approach required means that the different disciplinary specialists need to work together and mutually plan, execute and evaluate their activities -- thus tying together biophysical and socioeconomic facets. The optimal composition of the team will vary with circumstances, such as the land use problems of the area and the existing farm enterprises. However, ideally the team members should either represent the following

---

68 Whereas a watershed is a clearly recognisable natural hydrological unit, its boundries rarely conform to the cultural, administrative and political boundries in which individual farm-households actually operate. For this reason the new approach to soil conservation places greater emphasis on developing soil conservation strategies whithin the context of social/administrative boundries, rather than whithin the confines of natural watersheds [Douglas, 1993].

general areas or have ready access to individuals representing them, namely crop sciences, livestock sciences, silviculture/forest management, social sciences and natural sciences (i.e., including a soil conservation specialist). To address gender issues adequately it is highly desirable that at least one of the team members is female.

The time required to successfully implement the new approach to soil conservation will vary according to a variety of factors including:

- Size of the target area
- Complexity of the farming systems (e.g., range of component enterprises and number of recommendation domains)
- Nature of the problems
- The team's experience and knowledge of the area
- Suitability of existing land use recommendations
- Strength and experience of existing research and extension programmes
- Success in identifying relevant alternative improved farm management and land use practices

The time ideally required, and the time actually available to do the work, are not always the same and will depend on factors such as the perceived urgency of the problems and possible pressure from politicians and others to come up with quick answers. It will also be affected by the degree of governmental/departmental support that will determine the staff time and financial resources available to do the work. If at all possible, it is important for the team to indicate to the appropriate decision/policy makers an estimate of the time required to derive satisfactory results, so that expectations are not unrealistically high. Logistics are another important issue influencing time. Inadequate attention to factors such as availability of support staff, transport, office facilities, equipment and supplies, and financial resources, can affect not only smooth implementation of the exercise, but also the time it takes, and the potential pay-off from the work.

Finally, a key activity in the preparation is to establish contact with key institutions and individuals within the target area. This will include the local offices and representatives of the relevant public sector institutions, as well as locally operating NGOs. Contacts will also have to be made with local leaders, prominent individuals and groups within the community covering civil, religious and political matters. The reason for these initial contacts is to brief representatives, of those working and living in the area, as to the purpose and nature of the planned activities and to obtain their permission and support for its implementation.

5.4.2 Collection and Review of Secondary Data

For most areas in which small-scale farming takes place there will already be a wealth of information available on both the natural resources and the farming community. The justification for spending time and effort obtaining secondary information is that it results in a quicker understanding of farm household circumstances, the components of their farming systems and the biophysical and socioeconomic environments in which they operate. The importance of reviewing secondary information before actually starting any data collection at farm level, is that it avoids duplication of work. It also makes it easier to engage in discussions with local people. Farmers judge outsiders on the basis of their behaviour, attitudes and questions. Irrelevant and culturally insensitive questions can result in distrust and lack of cooperation. Therefore, if team members can demonstrate that they have a basic prior knowledge of the local issues of concern, this will help in establishing rapport with farmers.

There are often many sources of secondary information. Examples of types of information that are often available and can be particularly useful, include: agroclimatic information; topographic maps; aerial photographs; satellite imagery; land use, soils and other natural resource maps and reports; census reports; data and reports on health and nutrition; anthropological and social information; data and reports on irrigation, agricultural production, markets and prices; extension recommendations; experimental results and reports; and, information on the macro-economic and policy environment.

It is rarely appropriate and practical to try and collect all the information mentioned above. The need is to try and determine what information is really needed and whether a particular issue has relevance to developing an understanding, within the particular study area, of the different farm household circumstances and their production and sustainability problems. The type of information as mentioned above is not complete and in any specific situation the information required may be different from those mentioned. Similarly the source of information on a particular topic may vary from area to area. Arising out of this exercise it should be possible not only to obtain some understanding of the characteristics, problems and potential of the target area, but also help in highlighting the major information gaps.

5.4.3 Field Reconnaissance

During, or immediately following, the collection and review of secondary data, it is often useful for the team to make a two to three day visit to the study area. The purpose is for the team to obtain a visual overview of the study area's physical environment and land use patterns and to be briefed by locally knowledgeable people. In this way they can develop an initial feel for the area's specific circumstances and put the secondary data reviewed, into its context. Two key activities should be undertaken during the initial field visit, namely a reconnaissance tour of representative parts of the study area, and initial consultations with local key informants.

The reconnaissance tour can often be conducted from a vehicle and consist of a 'windshield survey'. It should however include frequent stops to allow the team members to view the landscape, to visit representative land management units, have informal discussions with local residents, and in general begin to develop a 'feel' for the area. To see examples of all the land management units and the different farming systems it may be necessary to undertake part of the tour on foot. This is to ensure the team visits areas representative of those cultivated by remote and isolated communities in the parts of the study area not accessible by road. It is advisable for the team members to be accompanied during the reconnaissance survey by local guides and researchers or technicians familiar with the area. These persons may themselves be sources of information on the area, and may also serve as intermediaries, assisting the team to obtain information directly from local farmers and qualified informants.

During this field visit initial consultations with local key informants can provide a great deal of information. Such consultations will also provide a timely opportunity for clarifying and supplementing information obtained from the desk top review of secondary data. Key informants will be people particularly knowledgable about one or more aspects of the local situation. They may include local government officials, teachers, extension workers, veterinary assistants, rural health and community workers, village headmen, innovative farmers etc. as well as research workers and land use planners familiar with the area. When it comes to obtaining information on farmers' attitudes to particular problems and recommendations, this may be the only source of information available prior to carrying out household interviews.

5.4.4 *Preliminary Characterisation of Farm Household Circumstances*

By now the team should have sufficient background information to have developed a preliminary appreciation of the study area's specific circumstances and the dominant production and sustainability problems of its farming systems. Although a more detailed assessment will be undertaken later it is worthwhile for the team to use the knowledge gained so far to prepare a brief characterisation of the farm household circumstances. Types of information that can be useful include: history of the area and its cultural characteristics including community organisation and structure; basic statistics and trends; land tenure and user rights; the natural resources; crops (i.e., annuals and perennials), livestock, trees; food production and availability; soil and water conservation; and, support systems, and national and local policies.

As a result of this exercise, the team should be able to identify and define, on a preliminary basis, different farm household groups or recommendation domains. A brief description should be produced of each recommendation domain detailing the household resources available, the land use enterprises pursued, and the constraints, problems and potentials of the farming systems.

A good preliminary assessment will save time later by focusing the team's attention on key characteristics of the area and potential land use problems and potentials requiring subsequent investigation and verification. However, the team should be aware that it is a preliminary assessment, and as more information becomes available in the course of undertaking the next steps, it may be necessary for them to alter and drastically revise their initial interpretations of the farming circumstances and land use problems.

This step also provides an opportunity to identify gaps in the information available to the team. Knowing where the gaps are, will help in undertaking the final activity in this step, which is to prepare the terms of reference for the part of the exercise that involves primary data collection. With respect to this, the team will therefore then need to determine what needs to be done, who will do it, which techniques and procedures should be used, at what level of detail is required, what modifications may be necessary to adapt specific procedures to the situation in question, etc.

5.4.5 Primary Data Collection to Assess Household Circumstances and Problem Identification: Informal Exploratory Survey and Analysis

Table 4 implies that primary data is first collected and then analysed to determine or confirm the relevant recommendation domains, and identify the constraints and the appropriate intervention points. In reality, these activities are implemented simultaneously. Initial data collection, particularly when done in an informal survey format, raises issues and hypotheses that are then confirmed or adjusted in the light of later interviews.

Not surprisingly therefore, we have already discussed the desirability of trying to collect the necessary data during the descriptive/diagnostic stage via an informal survey format. If necessary this exploratory descriptive/diagnostic activity can later be supplemented by some form of verification activity -- which can, if absolutely necessary, be more formal in nature. However, we want to emphasise, once again that, whenever possible, most emphasis and reliance should be placed on informal survey techniques because:

- They provide the best opportunities for interacting with farmers and for collecting and usefully using the knowledge they possess.

- The approach provides a time and cost efficient way of collecting information that enables the team to move on to the next phases involved in implementing the new approach to soil conservation.

Consequently, these informal surveys have four important characteristics that are particularly helpful at this stage:

- There is direct technician-interviewee (i.e., usually farmer) interaction, because the interviews are conducted by the senior team members themselves.

- Interviews are basically unstructured and semi-directed (i.e., semi-structured). Questionnaires are not used, although teams often use topic guidelines to make sure all relevant topics on a given subject are covered. Sometimes, written notes are taken during the interviews, with the emphasis on dialogue and questioning to provide information.

- The data collection process is dynamic and iterative because researchers evaluate the data collected and reformulate data needs on a daily basis. Interviews often first cover the broad characteristics of the farming system. Later, they focus on priority problems, potential solutions, and the interactions of these with aspects of the system.

- The interviews are conducted by an interdisciplinary team, with each discipline contributing collectively to the identification of problems and identifying and evaluating potential solutions. In doing this, they take into account factors influencing both the biophysical and socioeconomic environment within which farmers operate.

Some of the techniques used in undertaking these activities are discussed below. Obviously not all the techniques are ever used in any given situation. The aim is to indicate the variety of approaches that exist for collecting and analyzing information and data that will help in initially identifying the problems and then help in proposing relevant solutions to those problems.

Examples of interviewing techniques that can be useful at this stage include the following:

- ***Key Informants***. These are individuals with a special knowledge of the topic of interest (e.g. respected leaders, chairperson of the forest committee, etc.). They are major sources of information for those conducting in-depth research or for those interviewing under time constraints [Molnar, 1989].

- ***Group Discussions.*** Group discussions with a spontaneous, non-representative assembly of farmers can yield quick understanding of farm-household situations. They can provide useful information on a wide range of topics including, for example:

  - Normative information (e.g., average yield) and variability.
  - Features common to all farmers (e.g., marketing channels, prices, general constraints).
  - Community organization and power structure.
  - Local units of measurement.
  - Changes in farming practices over time, perceptions of soil erosion, etc.[69]

A special form of group discussion is called focus-group interviewing. Here, the group is deliberately chosen (e.g., a certain age group, smallholders growing fodder trees, or women household heads, etc.). When smaller, homogenous groups of people are naturally assembled during the field visits, this type of interviewing is appropriate to gain in-depth information on particular issues. It is also used to obtain a consensus of opinions from a group of 'local professionals' (e.g., local foresters, extension personnel). Questions in these interviews are highly focused on a few key issues. This type of interview is generally not useful at an early exploratory stage, before the general features of the farm-household system are understood [Molnar, 1989].

[69] PRA techniques, such as getting groups of farmers to draw maps, construct trend lines, do matrix ranking and scoring can be particularly useful in eliciting this type of information.

- ***Iterative Discussions.*** Iterative discussion with farmers is a form of communication that allows time for reflection, both for villagers and the team. One possible structure is to have extension agents, villagers, and technical specialists engage in a series of discussions. First, the villagers meet with everyone as a whole for an introductory session. Then the villagers are divided into separate, homogenous groups -- or assembled at a separate time -- which discuss the same issues from the perspective of their own needs and concerns. Then the village as a whole is reassembled to discuss the issues. At this point, the less outspoken village members, who have had a separate chance to speak apart from other villagers, are more willing to air their differences and add perspectives. Several such sessions may be needed before the process is complete.

On occasion, more direct means of collecting data may be desirable, as long as they do not take too long and another, more indirect or cheaper method cannot be used. Four particularly useful methods are as follows:

- ***Topographic Transects.*** These are longitudinal or transverse sections of a particular area that show the relationship between the use and the characteristics of each section of land. They are particularly useful where there is a strong relationship between land use and topography, such as in mountainous areas. Whenever possible such transects should be undertaken as a participatory exercise with the team and farmers walking the transect together. Such transects are useful in discussions with villagers to clarify the variety in land conditions and the corresponding differences in land management [Mascarenhas et al, 1991]. A series of transects can be used to show major shifts in land use or land management over time.

- ***Village Mapping.*** A technique that is currently becoming very popular, involves asking a representative group of villagers to help in constructing a map,[70] often by drawing large-scale on the ground, the main features of the village and its surrounding areas (e.g., mosques and other main buildings, residences having families with specific characteristics, major roads and rivers, hills, major farming and grazing areas, major soil types, and areas of degradation). This information can then be recorded permanently on paper. This approach can be very helpful in giving information about the village, and in certain situations, could provide a useful way of drawing up a sampling frame for use in selecting farmers/fields for trials or more formal surveys.

- ***Direct Observation***. The team, while moving through the project area, observes the farmers' fields, practices, circumstances, homes, and household activities. The members' observations can be used for checking the validity of information obtained through farmer interviews. In this regard, the experience of individual members will be critical in identifying the underlying causes of observable and stated problems. Such information often will suggest follow-up questions to reconcile any inconsistencies between farmers' statements and the team's observations [Douglas, 1988]. Through direct observation, major forms of land utilization may become apparent. Buildings and other assets can serve as indicators of the wealth of a family. Nevertheless, caution needs to be exercised, because direct observation relates to one point in time and may not reflect a valid picture over a longer time perspective.

- ***Learning by Working***. Physically working with farmers and doing what they do can greatly facilitate the exchange of information between technical specialists and farmers. As a result, it can lead to very meaningful collaboration. Working at farm tasks with farmers in their

[70] Details on how to conduct village mapping as well as a range of other useful participatory appraisal techniques can be found in Mascarenhas et al [1991].

fields will often elicit details the researcher would not normally consider. Most farmers generally take their role as teacher very conscientiously [Chambers, 1983].

A number of other available techniques not only help ensure the collection of relevant data, but also help in its interpretation in such a way as to indicate the major characteristics and constraints in the farming systems being practised. Some of these are as follows:

- ***Glossary of Local Terms.*** Compiling glossaries of local terms and concepts not only leads to improved communication, but often provides information vital to an understanding of local environmental conditions. The range of subjects that can be covered depends upon needs, but can include colours, climate, time, soils, plants, topography, micro-environments, animals, insects, foods and diets, diseases, cures, pests, weeds, seasons, space, measurements, proverbs, social relations, and ceremonies. The insights obtained increase understanding of local beliefs and practices. They also improve communication through enabling staff to use categories and expressions that are locally familiar [Chambers, 1983].

- ***Minimum Data Sets.*** Minimum data sets are lists of essential aspects or indicators concerning a certain topic. These 'checklists' provide a framework for collecting information about that topic. Minimum data sets are developed to ensure that important aspects of a particular issue are covered. Well-defined data sets ensure that information irrelevant to the focused questions being asked is not collected to the detriment of the problems in need of analysis [Molnar, 1989]. These types of 'checklists' are particularly helpful when the interviewing team members are relatively inexperienced and need some guidance on the topics to be covered.

- ***Learning about the Informal Research and Development System (i.e., Indigenous Knowledge).*** In agricultural communities there exists a dynamic and productive informal research-and-development (R and D) system [McCorkle, 1989]. This system interacts with new technologies or organisational structures introduced from outside. Unfortunately, in the past, the importance of research and developmental activities by farmers frequently has been overlooked. However, their knowledge and contributions are now increasingly regarded as being critically important in developing solutions to the problems that are identified [Haverkoort et al, 1991]. Three interrelated types of information coming from informal R and D systems are [Biggs, 1980]:

    - Technical and organizational innovations that make efficient use of resources scarce to farmers.
    - Ideas for new research on which scientists in the formal R and D system might start to work.
    - Methods for conducting cost-effective research and classifying knowledge.

- ***Ranking Techniques.*** A wide range of ranking techniques can be used to explore with farmers their perceptions of a particular topic, to elicit values and constructs, and to discover indigenous views of the world. These techniques are particularly valuable for demonstrating how the perceptions and beliefs of outsiders and informants are often radically different. Examples of the uses of ranking techniques include the following [Pretty, 1989]:

    - Preference ranking: to make a ranked list from most preferable to least, with favourable and unfavourable criteria indicated for each item.

- Comparative ranking: to gather information quickly on seasonal or yearly patterns, indicating when a topic occurs most and least often.

- Direct matrix ranking: to rank several types of a certain item (e.g., soil types, crop varieties, etc.) against a number of common positive and negative characteristics.

- Wealth ranking: to produce a ranked list of households according to wealth as perceived by the informants.

- Problem ranking: to produce a ranked list of problems and to generate a discussion about priorities.

- ***Causal Diagramming***. This is a technique for determining and explaining the complex relationship between different causes and how their effects, individually or combined, lead to a particular problem. The various causal factors related to a problem can be linked into different causal chains. The causal diagram can be used to identify possible intervention points. The team, with the help of the farmers, then determines which have the greatest potential for solving the problem. It also can indicate what interventions are necessary before other potential interventions can have any effect [Douglas, 1989].

- ***Decision Modelling***. The reasons why farmers use certain farm practices or take certain household decisions are often complex and not readily apparent to outsiders. Modelling of farmer decisions can provide a better understanding of the decision criteria and can also indicate certain 'leverage points'. These are points in the farming system where there is scope for introducing improvements in ways that are likely to be feasible and acceptable to farmers. By interviewing a number of farmers, either informally or formally, the decision criteria used for a particular decision can be identified. These criteria are then arranged in a flow chart in a logical manner, for example, in the form of a decision tree. By indicating the number of farmers that use each decision criterion, the decision model shows the relative importance of criteria in a particular decision. Furthermore, the model can be used to show how farmers explicitly weigh the relative importance of any two particular criteria [Franzel, 1984].

- ***Historical Profile***. A historical profile is a technique to view current problems within an historical perspective. Such a profile describes, for example, what major factors have directly or indirectly influenced land use in the area over time. These could include settlement history, increase in population levels, introduction of particular crop or livestock enterprises and practices, fluctuations in market prices for important crops or inputs, or colonial history. These factors can be plotted from a relevant time period to the present, and compared with a similar historical profile that records when particular land use problems were first observed, and any changes in their degree or severity since then [ICRAF, 1983, Douglas, 1988].

- ***Seasonal Calendar***. A seasonal calendar depicts those factors that are important at different times of the year in the production cycle and compares the factors to detect constraints and potentials. For instance, different land management activities can be highlighted and compared with seasonal profiles of labour needs and availability. Similarly, so-called crop and livestock calendars can be drawn up, indicating major activities and stages in production. These can then be compared with seasonal profiles of climatic variations, ground cover, or erosion intensity [Chambers et al, 1981].

- ***Learning from Success***. The analysis of success can give very interesting and useful clues on working methods, production mechanisms, or survival strategies. Success here refers to the better than average results in farming or surviving, as compared to fellow group members in similar circumstances. The 'head and tail' technique is an example. A group of land users in similar circumstances is divided into three groups: the 15% (or 25%) most successful, as measured, for example, through net income or yields of a certain commodity; the 15% (or 25%) least successful; and the remaining average group. The management practices of the three groups are then compared, and each variable is analyzed in order to discern significant differences. Then proposals in the design stage may be based on the experiences and methodologies of the successful farmers [Zandstra et al, 1979].

5.4.6 Primary Data Collection to Assess Household Circumstances and Problem Identification: Verification Activities

Verification activities can involve studies and direct measurement activities as well as surveys. The verification survey is designed to confirm, in a more systematic statistical manner, the results arising out of the exploratory diagnosis. At the risk of being tedious, we wish to emphasise that quantitative primary data collection with respect to biophysical and socioeconomic characteristics can be very time consuming and expensive. Therefore, it should only be done if there is absolutely no alternative and there is a very real danger that inappropriate or irrelevant solutions to the identified constraints will be agreed upon and implemented. In a sense this should rarely arise if there is a genuine belief amongst team members that farmers possess an intimate and intuitive grasp of their biophysical and socioeconomic environment. Therefore if given the facts or information about potential solutions, they should be able to indicate whether or not they are likely to be relevant.

However, if it is unlikely that solutions exist to the problems identified, and, for example, it is necessary to enlist the cooperation of other individuals and/or institutions, then it may be necessary to use a more formal approach involving the gathering of more quantitative statistical types of information. Such information may be required, for example, in convincing researchers on experiment stations to work on a particular problem or to convince planners to adjust policy/support programmes. An alternative approach, if such an eventuality is likely to arise, is to encourage influential representatives of such institutions to participate directly in the exploratory descriptive/diagnostic activities. Scientists and planners are, like most people, seldom immune to interaction with, and in-depth discussions with others, including farmers!

If, in spite of these efforts, more formal studies or surveys are required, they need to be done in a sufficiently systematic manner to convince the interested parties or potential collaborators, in a manner that is statistically acceptable. For example, in terms of natural resources, a systematic soil sampling procedure may need to be implemented to collect soil samples that can be subjected to laboratory analysis and perhaps later correlated with farmers' own classification system.

In terms of more formal socioeconomic surveys it is necessary to select samples according to a generally accepted sampling procedure. For example, some form of stratified random sampling is often preferred, with strata being the farming recommendation domains. To save time, these surveys usually consist of one visit per respondent, using a formal structured questionnaire dominated by close-ended questions. Care is taken that excessive amounts of data are not being collected and that the survey questions are carefully worded. A critical component in this exercise is to maximize the efficiency of the data collection, processing, and analysis link so that the results are available in a timely manner. The quantitative data emerging from these surveys can be analyzed in many ways other than just by recommendation domain (e.g., by focusing on one, or more, individual characteristics such as owner versus tenant farmers, farmers using specific improved practices, farm size, gender of household head, etc.). Such analysis can help provide verify constraints and, with the help of information from the more informal surveys, and data from the formal natural resource surveys, help verify the direction of designing and testing/implementation activities.

For example, formal studies and surveys, when undertaken, provide an opportunity for quantitative analysis to be undertaken, against which future proposals and solutions can be compared in what is often viewed as a completely objective manner. For example, soil chemical analyses or definitive soil erosion measurements can help compare the results before, and after, an intervention while similar analyses in the form of economic and financial analyses can be done with respect to the socioeconomic area.

5.4.7 Analysis and Description

As we emphasised earlier much of the information relating to farm household description and constraint analysis emerges during the implementation of the informal exploratory/diagnostic survey. However, there is value at the end of this to pause for reflection, and a systematic recording and summarising of the results and impressions. This pause becomes particularly important if formal verification activities are undertaken, the results of which need to analyzed in a more formal manner.

Such analysis should provide confirmation of the major recommendation domains, with a description of their characteristics, their problems and constraints, and related causal factors. This helps in identifying potential intervention points, on which future activities need to be focused in order to improve productivity of the farming systems in a manner that is sustainable in the long run. This analysis or assessment process includes constraints and potentials relating to both the biophysical and socioeconomic environment.

In terms of the biophysical environment this will involve assessing the characteristics of the land and comparing these with the biophysical requirements of the various land use enterprises that comprise the farming systems of each recommendation domain. The aim is to determine how environmentally suitable the enterprises are for the area and specifically whether some of the constraints limiting farmers' production are due to natural causes. Also some of the land degradation may be a result of inappropriate land uses or management practices, given the biophysical conditions present. Any natural constraints thus identified will have to be taken into consideration when developing improved conservation effective farming systems.

Techniques have been developed for evaluating the suitability of land according to its biophysical characteristics, and for comparing these with the requirements for sustainable production from existing and potential uses. The most widely used approach is that of land suitability evaluation developed by FAO.[71] Many of these concepts and procedures are considered applicable to the new approach to soil conservation. With regard to the analysis of the biophysical constraints the important feature of the FAO land evaluation approach is that it is based on two parallel sets of activities. One involves identifying and describing the characteristics of the land itself, the other involves describing the kinds of land use and assessing their biophysical requirements. By this point, it should be possible to divide the target area into discrete relatively homogeneous areas or land units. As well as mapping the boundaries of these land units the team should have been able to produce detailed descriptions of the biophysical properties of each. The properties of each land unit can then be considered in terms of their:

---

71 Detailed guidelines on the approach can be found in a number of references, namely FAO [1976, 1983, 1984, 1985, 1988A].

- ***Land Characteristics***: these are attributes of the land which can be measured or estimated and used for distinguishing between land units of differing suitabilities for use. They are employed as a means of describing land qualities. Examples are: mean annual rainfall, slope angle, soil drainage class, effective depth, topsoil texture, soil available water capacity, pH and soil nitrogen percentage.

- ***Land Quality***: this is a complex attribute of the land which acts in a distinct manner in its influence on the suitability of land for a specific kind of use. Land qualities may be expressed in a positive or negative way. Examples are: moisture availability, erosion resistance, flood hazard, nutritive value of pastures, accessibility. Where data are available, aggregate land qualities may also be employed, for example, crop yields, mean annual increments of timber species.

One or more, usually several, land characteristics will be combined in order to describe a particular land quality and the comparison of land with land use is usually done in terms of land qualities. If land characteristics are employed directly in evaluation, problems arise from the interaction between different characteristics For example, the hazard of soil erosion is determined not by slope angle alone but by the interaction between slope angle, slope length, permeability, soil structure, rainfall intensity etc. Because of this problem of interaction, it is usual for the comparison of land with land use to be carried out in terms of land qualities.

One of the principles of the FAO framework for land evaluation is that land suitability is assessed and classified with respect to specified kinds of use. By matching the biophysical requirements for the different enterprises with the qualities of a particular land unit, it becomes possible to assess their suitability for the different land management units within the target area. This needs to be an interdisciplinary team effort as it integrates knowledge of agronomy, animal husbandry, range management, silviculture and natural resources (i.e., climate, soils, topography, vegetation, etc.). In determining the requirements for a specific crop, livestock or tree based enterprise, the team may need to supplement their own knowledge by contacting specialist sources.

Once the relevant land qualities of the land management units and the requirements of the land use enterprises have been determined, they are compared to see how well they match. The aim is to answer the question "How well do the qualities of the land satisfy the requirements of the land use enterprise?" It is usual to classify a particular land use as being highly suitable, moderately suitable, marginally suitable or not suitable for a particular land management unit. By looking at the suitability ratings for all the land use enterprises within a recommendation domain, it will be possible to determine the suitability of the typical farming system according to the natural circumstances facing the farmer. Ratings of land suitability, other than highly suitable, indicate that the land qualities do not match the land use requirements for optimum production. By identifying which land qualities limit production and increase the risk of land degradation, and by determining how far they down grade the suitability rating, it is possible to determine the type and severity of the natural constraints.

Once identified, the team will need to decide whether it may be possible to develop means of overcoming them so as to permit a sustainable increase in production, or whether it is necessary to regard them as fixed constraints (i.e., cannot readily be changed). In the former case the need may be for improvements or changes in the existing management practices or component technologies (e.g., the introduction of contour hedgerows/grass strips into arable cropping on hill slopes or drought resistant cultivars in drought prone areas. In the latter case the team will need to find alternative land use enterprises that are not only more suited to the qualities of the land but which also conform to the goals and objectives of the existing farm household systems.

Whereas the emphasis in the above description of land suitability evaluation has been on matching land qualities and land use requirements it is common to include socioeconomic considerations in the evaluation. This involves not only the biophysical suitability of a specific crop, livestock or tree enterprise for a particular area, but also its social, economic and financial suitability. In the context of the new approach to soil conservation, this calls for integrating the biophysical and socioeconomic assessments, to determine whether the returns that can be expected from existing or potential land use enterprises are acceptable to the farm household, given the costs of production and the biophysical potential of the land. A low potential yield of a high value crop may still be financially attractive to a household, and considered suitable for inclusion in the farming system, if alternative crops, whose requirements better match the land qualities, have a lower market value.[72]

A number of other types of analysis, particularly those relating to the socioeconomic area are very important in determining the current problems, constraints and potential for developing alternatives that are attractive. In this regard, two types of analyses that are particularly helpful in developing or confirming insights. These are:

- ***Economic and Financial Analysis***. In economic and financial analysis, benefits and costs of activities are estimated. They are expressed in terms of money, so that calculations can be made. In economic analysis prices are used that reflect those to the society as a whole, while in financial analysis they reflect those paid, or received, by the farmers themselves (i.e., including subsidies, etc.). Various types of analyses can

---

72 An alternative approach to that of land suitability is land capability. However, for various reasons discussed elsewhere [Douglas, 1992], we do not consider this such an appropriate method.

be undertaken. For example, an analysis can be made for the activities of a typical household or farm (e.g., representative of a given recommendation domain) or for a particular crop or livestock enterprise. Gross margin analysis is a technique commonly used for this purpose.

- ***Cash Flow Analysis***. Cash flow analysis provides information about the cash flow within the household: income in the form of cash from sales; credit and loans; off-farm revenues; and costs in the form of purchases, investments, interest, taxes, marketing costs, etc. The financial information can be used in a benefit/cost analysis in which there is a stream of costs and benefits over a number of years (e.g., as a result of a specific soil conservation measure that pays-off in the long run). Whether for a short or long period such analysis can be very useful in explain some of the household needs and constraints.

However, socioeconomic analysis should also consider other facets such as:

- Some idea of the levels and seasonal use of resources (i.e., particularly labour) and seasonal variation in availability and consumption of food.

- Some welfare analysis including determining whether intrahousehold relationships are equitable in terms of access to food, obligations and rewards, whether they have sustainable means of livelihood in the long-run, whether they can meet the social and cultural obligations of the community, etc.

These types of analysis can help provide additional information helpful in determining problems and constraints and in so doing give some indicators on what might be suitable interventions -- the focus of the design phase.

Having said that, it is important to stress that the identification of problems and opportunities is an iterative and dynamic process that continues indefinitely. Also the new approach to soil conservation is not concerned with overcoming all the problems faced by farm households but is specifically concerned with those that affect their farming systems' productivity and sustainability. These problems are generally the result of biophysical and socioeconomic constraints that prevent households from attaining their production goals and targets. In order to tackle these problems it is necessary to identify the causes and any constraints that have to be overcome to solve them. In addition to being faced by problems, farm households will also have opportunities, that is potential options for relieving constraints, overcoming problems and increasing production. The identification of problems and opportunities therefore go hand in hand.

In the more traditional approach to soil conservation, it has been 'outside' development experts who have assumed responsibility for identifying the problems and opportunities of farm households prior to designing conservation and agricultural development programmes for implementation at the local level. This has often led to failure because the priorities and objectives of outside experts and local farmers are likely to be different. Hence they can be expected to have different perceptions as to what the real problems are. A programme designed by outsiders to tackle an issue, such as soil erosion, that is not perceived by the target beneficiaries as a problem, is unlikely to gain their support and voluntary participation. We hope by now, that it is obvious that we believe it is critically important to involve the people of the target area in the problem identification process and in the ranking of their importance. Much of this can be done through the informal survey process and more formal approaches that we have discussed. The team is likely by now to have developed its own list of problems and possible intervention points. This list is likely to include problems not only relevant to farming households in different recommendation domains, but also to the community and society as a whole. Farming families themselves, because of operating barely above the survival level, may understandably have a narrower perspective. The lists of problems and ranking of their importance drawn up by individual farmers and groups of farmers should ideally be confirmed at village level participatory workshops and compared with those derived by the team. It is particularly important in such circumstances to ensure that opinions of all factions of the community are canvassed (e.g., women and poorer households as well as the leaders).

In addition to a wider societal perspective, the team may sometimes identify problems not recognized by farmers but for which opportunities for their alleviation exist. These opportunities may be the better use of under utilised or misused resources or the introduction of new technologies. However, as we will emphasise in the section on the design phase (i.e., Section 5.5), we believe at the outset it will be better to target problems perceived by farmers as priorities and then later move on to informing them about other problems and opportunities once their confidence has been gained. In some cases it may be possible to solve one problem as a result of solutions to another (e.g., tackling a livestock feed shortage, recognised by farmers as a problem, through introducing contour strips of perennial fodder crops into arable fields, thereby controlling surface runoff and erosion under annual crops, which was perceived as a problem by the team but not by the farmers).

At this point, we should perhaps briefly mention the issue of looking at problems from a farming household or societal perspective. If the problem is of major concern to both the local farm households and to society at large then there will be mutual agreement to it being tackled. Where the interests of the farm household and society diverge, two possibilities exist. If the problem is of interest to the farm household, but not a concern of society at large, then it becomes difficult to justify the team spending much time and money on trying to solve it, given the need to fit its activities into the overall development objectives and policy of the government. Alternatively if the problem is in the interests of society, but not a concern of the local farm households, the government has the choice of whether to alter the farming environment, including incentives, to bring the farm households' interest in line with those of society, or to leave conditions as they are.

At the end of this exercise when analysing farm household circumstances, the major production problems should have been determined and the major constraints to overcoming them should have been ranked in terms of importance. Also, the sustainability problems and constraints facing farmers, from both existing and potential land degradation, will have to have been identified. As well as identifying existing land degradation problems the team will need to predict likely future trends and therefore possible future problems. Unless the causes of a problem are clearly understood inappropriate interventions may arise resulting in the wrong recommendations being made to farmers. Causal diagramming, discussed earlier, is a valuable technique for determining and explaining the complex relationships between different causes and how their effects, individually or combined, lead to a particular problem. By clearly demonstrating these interrelationships the technique can help avoid or resolve differences between team members as to which of their list of problems are the key ones, which are in reality causes, and how the various causes and constraints identified are linked. It can also show to the team that, should they seek to remove one cause or overcome a particular constraint without dealing with the others that contribute to the problem, their proposed interventions may have little overall impact.

Therefore, at the conclusion of the data collection and analysis activities, there should be for each recommendation domain, an understanding of the characteristics of the farm households, the problems and constraints they face, some ranking of the importance of those constraints, and some notion of the potential skills and resources they have at their disposal to address those constraints. At the same time, there should be a similar understanding of the problems and constraints relating to production and sustainability that go beyond the farm gate and involve the community or society as a whole.

5.4.8 Broad Scale Considerations for Development of Solutions

Having identified the problems and decided where there is scope for making interventions, the final step in the data collection and analysis stage of the new approach to soil conservation is to determine the broad technical and policy issues that need to be considered when developing improved land use enterprises and farm management practices for dissemination to the farm households of the study area. In this respect some of the issues which will need to be considered are:

- The team will need to determine whether there is scope for simultaneously tackling the production and sustainability problems jointly, or whether the causes and solutions to these problems are so different that they need to be dealt with separately. Obviously, as we have stressed earlier, it is most desirable to develop strategies that not only offer short-term production benefits but that are also conservation effective (i.e., conservation effective production technologies). This is particularly the case in those areas where land degradation has not yet become severe, leaving scope for adopting a conservation strategy based on prevention rather than cure. Hence the primary thrust should be to focus on developing and disseminating new and improved production oriented land use enterprises and farm management practices (i.e., improved land husbandry practices). However, if parts of the study area are already severely degraded, the conservation problems may be so severe that a more conventional rehabilitation strategy is required (i.e., corrective approach based on proper understanding of the causes). In such a situation there may be little scope for tackling production problems until the present misuse of the land has been corrected and a rehabilitation programme implemented to restore the land to a productive condition. The emphasis may be on the use of physical structures to prevent further erosion, and to use machinery or labour to fill in gullies, construct terraces, etc.[73] Hence one of the first issues for the team to decide is whether the strategy adopted, to tackle the sustainability problems, should follow a preventative or corrective thrust. Hopefully, it is usually the former.

- It will be necessary to decide, before a lot of effort is put into designing technical solutions, whether there is an equal or perhaps greater need to make policy changes. This may be to remove policy constraints that prevent the adoption of productive and sustainable farm practices or to enact new policies that will facilitate adoption. This may also require determining whether there is a need to eliminate conflicts between development policies designed to promote production in the short-term and those designed to conserve the environment for use by future generations. There may be little point developing improved field level technologies if external constraints within the policy environment remain unchanged. For instance farmers may be unwilling to plant trees in their farm holdings until there is a change in government policy on the legal ownership and protection status of trees in farmland. Likewise there may be a need for a change in the government guaranteed price for a particular crop before farmers will grow it.

- A key concern in the subsequent activities is should the problems be tackled, at the farm, community or a higher level? As a general rule the new approach to soil conservation implies that, whenever possible, problems have to be tackled at the farm level and solutions implemented by the farm household members themselves. Larger-than-farm scale problems have of necessity to be tackled at the community level or even at a higher level.

---

73 We, of course, recognise that sometimes less drastic means of overcoming such problems, are sometimes possible (e.g., see discussion on vegetative techniques in the footnote at the end of Section 3.4).

- A number of other difficult decisions also need to be made in laying down the framework for the design phase activities such as:

    - Are some of the constraints so binding (e.g., very severe land degradation) that the few resources available to the team, community and the nation, would be better spent where there is a better chance of some pay-off in the near to mid-term?

    - Is there scope or need for improved land husbandry practices for the farming systems as a whole, or just for part of it, and can this be done at the individual household level or does it need to be done at a higher level (e.g., improved productivity and sustainability of communal grazing land)?

    - Can the problem can be tackled within the resources of the farm household or local community or will it require external interventions?

    - Is there scope for making adaptive improvements to farmers' existing land use enterprises and farm management practices to make them more sustainable, or is it necessary to develop totally new ones? The former strategy is likely to be more acceptable to farmers.

    - Should future activities focus solely on overcoming the identified problems within the existing farming systems, or should they also seek new opportunities that would tap the inherent potential within them, thereby permitting optimal use of the resources available to each household? Farmers tend initially to respond more favourably to recommendations that offer solutions to what they perceive as problems and which enable them to intensify existing production.

    - Is there a possibility of phasing in interventions and leaving some flexibility to leave scope for future developments? Related to this is whether there is potential for a range of options for farmers rather than a single package.

The aim of this step in the analysis process is to complete the groundwork and determine the context for the next phase of applying the new approach to soil conservation, namely the design of technical and policy solutions.

5.5 THE DESIGN PHASE

The problems identified and ranked in the descriptive/diagnostic stage provide the background against which potential solutions are developed during the design stage. In Table 4, four steps (i.e., B1.1 to B1.4) are suggested for accomplishing this activity. These are briefly considered in the following sub-sections.

5.5.1 Review of Land Use Enterprise/Farm Management Options

Before formulating detailed solutions to production and sustainability problems it is important that there is a thorough review of the existing options for land use enterprises and farm management practices (i.e., farmers present practices and current extension recommendations) as well as potential alternatives. It is unproductive to seek to develop a package of improved practices without reference to the body of existing technological and farmers' indigenous knowledge. It is critically important that this review does not restrict itself to looking solely at the conventional soil and water conservation measures with its emphasis on physical structure. Incorporating the conservation dimension into FSD requires that this review should be broad ranging and cover not only soil and water conservation (i.e., both physical and biological practices), but also crops, livestock, trees, etc., and their related management practices.

In deciding among the alternative possibilities it will be necessary to know what land use options are currently available locally and the management practices associated with them. In reviewing the existing options the team will need to determine how well suited they are to the biophysical and socioeconomic circumstances of the farm households. If currently recommended options are suited to the biophysical properties (i.e., the land use requirements match the land qualities), but are not currently adopted, can any socioeconomic constraints be overcome through appropriate interventions?

Farmers' reasons for either following or not following a particular recommendation will be useful in reviewing existing options and currently recommended farm management practices. These can then be usefully used in deciding on possible solutions which will most likely involve exploiting positive relationships between some of the options (e.g., integration of crops, livestock and trees through alley cropping).

As we have repeatedly implied, farmers' land use practices are strongly influenced by the policy environment in which they operate and it is therefore important to look not only at the field level technical options but also potential options at the policy level. The team should first seek to identify the key elements of the existing policy environment and to review their effect on farm households' land use practices. For example, to what extent do government policies influence producer prices, markets and input distribution and does this affect a farm household's choice of land use enterprise? It will be important to distinguish between those influences that conform to the stated goals of government policies and those that relate to problems in implementing them, since the actual results of a particular policy may be quite different to the stated goals [Hudson, 1991]. This review should also involve looking at alternative policy options that could have a beneficial impact on the external socioeconomic circumstances and constraints influencing land use at the household level. An example would be changing the prices paid for particular crops, to encourage or discourage their production.

The common argument in favour of a policy of providing direct financial incentives for the construction of soil conservation measures is that, without them, farmers will not adopt the recommended practices. So, it is argued, for the sake of preserving future benefits for society, farmers should receive a short term incentive to compensate for the costs involved. However, we believe this should be avoided whenever possible. We would argue, that rather than use the incentive policy option involving adapting the farmers to the recommendation, it may be more cost-effective to find an alternative technical option, in that the recommendation should be adapted to the goals and circumstances of the farmers. An example would be a conservation effective production technology which yields short term production benefits while at the same time positively contributing to long run sustainability.

Where there is a need for external intervention to facilitate the adoption of particular recommendations (i.e., for both production and sustainability purposes) then there may be scope for policy interventions involving the use of indirect subsidies. An example would be selling tree seedlings to farmers at less than full production costs. However, any such indirect subsidy policy option should be either short term (i.e., only needed to overcome an immediate constraint to the initial adoption of the recommendation) or sustainable over the long term from revenue budget sources (i.e., to cover recurrent seasonal costs associated with the recommendation). The aim would be to make it easier for farmers to adopt conservation effective farming practices, perceived as beneficial in their own right, rather than as with the use of direct incentives, needing to 'bribe' farmers to conserve soil.

Another policy option, a favourite weapon of many bureaucracies, is legislation. There are many examples of conservation legislation on the statute books, which are not respected by farm households, cannot be enforced and therefore have little credibility. Instead of considering national level enforcement legislation it may be more productive to look at the options for formulating enabling legislation. This may be particularly necessary where a technical or policy option can not successfully be adopted without a change in the existing policy and legislative environment. For instance it may be necessary to consider introducing legislation that would provide, at the national level, the legal framework in which local communities can formulate and enforce their own by-laws. It is likely that there will be a much better chance of success through adopting a policy of promoting local land use by-laws, agreed on and policed by the local community. If these are formulated and implemented with the consensus agreement of the affected communities, they will always be more effective in preventing misuse of the land resources than any legislation passed at the national level.

5.5.2 *Selection of Potential Land Use Enterprises/Farm Management Options*

Following on from the review of options it will be necessary to identify and select those considered to have potential for combating the production and sustainability problems of the target area. The aim should be to quickly produce a list of potential options, including a range of possible alternatives. The options on the list can then be further reviewed to arrive at a shorter list of the most promising ones. It is these options that become the focus for the next step in the new approach to soil conservation involving detailed technical design and policy formulation.

Drawing up a reasonable list of relevant solutions is very dependent on effective interaction between all interested parties including farmers, team members, and representatives of any outside institutions (e.g., researchers, planners, etc.) who might need to be involved in developing and implementing the solutions. Four techniques that can be helpful in developing solutions, and a preliminary evaluation of their suitability, include the following:

- ***Brainstorming***. This is a technique intended to produce a wide range of ideas or suggestions, including imaginative and unconventional ideas about a specific topic. It is especially useful as a starting point for further discussion and review in situations for which the working strategy or a more restricted set of options has not yet been chosen. In situations that are new and relatively complex or in groups of people that have not yet had time to exchange ideas, this technique may give a first sense of order and indicate priorities and possible solutions.

- ***Workshops with Innovators***. Innovators are those farmers who are the first to try out new ideas. They are often isolated from each other. To capitalize on their knowledge and experience, they need to meet and discuss amongst themselves. In an innovators' workshop, suitable farmers are brought together to learn from each other and to generate ideas for further activities. 'Outsiders' are present to listen and learn and, where appropriate, contribute in an advisory capacity [Chambers and Jiggins, 1987]. Thus such workshops could potentially be very helpful in devising solutions to the problems that have been identified.

- ***Nominal Group Process***. This is a special-purpose process useful for situations in which individual judgements must be tapped and combined to arrive at decisions that cannot be made by one person. It is a problem-solving or idea-generating strategy in which participants indicate all their opinions about a given statement. Each group member then selects the ones that are felt to be most important and assigns scores to the preferred ones. The items that receive the highest scores then represent the group's priorities [World Bank, 1986]. This approach could be potentially useful in situations in which collective action is required in implementing a soil conservation programme (e.g., management of a watershed).

- ***Multi-Factor Decision Making***. This is a technique to evaluate alternative solutions for a given situation. Economic, ecological, administrative, and social feasibility are taken as decision criteria. First, for the given situation, the relative importance of each of these criteria is defined by ranking them on a scale from 10 (i.e., extremely important) to 0 (i.e., not important). Then, the different alternative solutions are evaluated by ranking them on four scales to indicate the seriousness of the economic, ecological, administrative, and social impacts. The values of the two rankings are multiplied and summed for each alternative. Finally, a matrix with the alternatives and decision criteria is drawn up to show the overall suitability of the different alternatives [World Bank, 1986].

The last two techniques can be particularly useful in a preliminary screening of the solutions which can be done by all involved parties, particularly farmers themselves and the team members. Hopefully in the end there is a reasonable congruency between the solutions preferred by both groups. If not, the team, at least initially until credibility and respect has been established with the farmers, should be biased towards investigating those solutions favoured by the farmers.

5.5.3 Development of Technical and Policy Action Proposals

Those most favoured solutions coming out of the exercise described in the previous sub-section are, at this point developed into detailed specifications. This means specifying in detail the technical requirements for each potential field level intervention and formulating the specific details of each policy recommendation. In the case of the former this would mean developing detailed research requirements or, if the necessary knowledge is available, drawing up extension recommendations advising farm households how to undertake specific improved farm enterprises and/or management practices. In the case of community level land use problems, the requirement would be for detailed guidelines on what needs to be done technically and how to implement it.

Although Table 4 implies that development of technical and policy action proposals, and appraisal are two discrete activities, in actual fact they are often undertaken simultaneously. The subsequent formal appraisal exercise is therefore simply a mechanism for checking that the appraisal criteria have been adequately considered in the development of specific plans of action.

In developing the technical action proposals, the following points need to be realised and taken into account:

- Different actions will be necessary for farmers in different recommendation domains.

- Although it may be desirable to devise an integrated 'model' farming system to exploit positive interactions in the farming system, it may be more practical initially to consider parts of the system separately. Nevertheless it is important to bear in mind one solution may solve more than one problem (e.g., planting a single tree species for both fuel wood and livestock fodder).

- A key activity in the development of improvements will be to determine the costs and benefits of the proposed improvements compared to the existing situation within the farm household system. Hence having identified technical options in which the 'land use requirements' match the 'land qualities', it is important to undertake an economic and financial analysis to ensure the potential returns to land, labour and capital are sufficiently attractive. The priority should be to identify low cost improvements that can be implemented at the farm level. However should they prove inadequate to tackle all the problems it may be possible to justify the use of high cost improvements.

The answers from the above should be used by the team to produce an integrated and flexible 'model' for an improved farming system for each recommendation domain which will best tackle the identified production and sustainability problems. The 'model's' component land use enterprises and farm management practices should ideally be:

- Simple -- be readily demonstrated to, and understood and implemented by, farmers.
- Low cost -- be within the financial reach of farmers, have limited labour requirements and require no foregone benefits (e.g., land taken out of production).
- Productive -- lead immediately to substantially increased benefits.
- Sustainable -- requiring limited effort or purchased inputs each year to maintain.
- Low risk -- not susceptible to climatic variations (i.e., drought or waterlogging) or market fluctuations (i.e., supply exceeding demand).
- Flexible -- leave scope for future developments (i.e., farmers can change maize varieties after one season but a decision to plant a long lived perennial tree crop is not so easily reversed).
- Conservation effective -- contribute to the maintenance of soil productivity (e.g., increase ground cover and soil organic matter levels, improve surface infiltration, reduce runoff, etc.).

The team should ensure that the final description is produced in a form which is understandable by, and agreed to, by all involved parties.

Given their position within the government hierarchy the team members will not themselves be able to directly change existing policies nor introduce new ones. However, the results of their analysis may reveal that present policies are contributing to the low productivity and land degradation of the study area. If so, they have a responsibility to identify areas where there is a need for policy changes and, to either advise on how existing policies could be modified, or to formulate recommendations for new ones. Such policy recommendations can then be forwarded to the relevant senior decision makers for their consideration and hopefully action. The fact that the team members cannot themselves change policies should not deter them from notifying those who can that there is a need to do so. If policy makers are not advised of the need for change they are unlikely to initiate change themselves.

5.5.4 Ex Ante Appraisal and Possible Revision

Having arrived at a series of technical and policy recommendations the team should undertake a formal preliminary appraisal exercise. This is a theoretical exercise undertaken before the technical options are tested or disseminated and the policy recommendations are passed on to senior decision makers.

The purpose of the appraisal is for the team to undertake a critical examination of the theoretical potential of the 'model' farming system(s), and the proposed policy recommendations, to solve the identified production and sustainability problems, and realise the potentials of the farm household systems of the study area, while proving acceptable at both the individual farm household and society levels.

The appraisal of the technical options should consist of the following elements:

- Is it technically and practically feasible to implement?
- Is it financially and economically viable and dependable?
- Is it sustainable not only in terms of the inputs required but also in terms of maintaining or preferably improving soil productivity in the long run
- Is it socially equitable and acceptable

In the past conservation recommendations have usually been judged according to the first criterion listed above, while those for increasing agricultural productivity have rarely considered more than the first two of the above. It is important to look at all the above in order to ensure that what is being recommended really does benefit the farmers it is intended for. Providing the 'model' farming system and its component enterprises and management practices can satisfy the above criteria, there will be a high probability that it will have the potential to make an impact on the identified problems.

Before moving on, perhaps a brief word on economic analysis is in order. For such analysis, we believe considerable reliance should be placed on relatively simple techniques. At this stage, of course, the data used are conjectural. In addition, two other critical issues underlying the legitimacy of many of the results of such analysis are:

- Placing correct values on the inputs and products that do not originate from or end up in the market place.
- The use of evaluation criteria that reflect those used by the household.

Particularly challenging is valuing labour -- which is likely to vary seasonally -- and evaluating the results in terms of the most limiting factor, for example, return per hour used during critical bottleneck periods, rather than return per hectare. Analytical techniques involving average return and budgets of various types (i.e., partial, enterprise, and sometimes whole farm) are commonly used. Marginal analysis -- a more powerful technique -- certainly is preferable where the technology is tested over a wide range of inputs.[74] Such a situation more likely occurs in RMRI type trials rather than at the RMFI and FMFI level, which are usually undertaken during the testing/implementation stage. Given the poor standards of living of many farming households in low income countries, dependability of return probably will be an important criterion for them in their assessment of the attractiveness of a technology. Yet, the greatest problems in easily applied techniques may exist in the area of risk.[75] Often, sensitivity analysis in conjunction with partial budgeting is the method used. Finally, because in soil conservation there are long-term benefits and possibly costs, some form of discounting the future is required. Benefit/cost analysis is used to accommodate this situation.

Turning to an assessment of the proposed policy actions, the appraisal may suggest to the team ways in which the policy recommendations can be modified to make them more acceptable to senior government officials and politicians and/or more likely to achieve the desired ends. The following criteria should be considered by the team in their preliminary appraisal of individual policy recommendations:

---

74 There are many texts available on these commonly used techniques. Hence there is little value in spending time in this publication discussing them in detail. Useful overviews concerning different parts of economic analysis exist in Dillon and Hardaker [1980], Worman et al [1990] and Douglas [1992]. CIMMYT [1988] provides a clear description of marginal analysis using on-farm trial data.

75 Techniques such as stochastic dominance, minimum returns analysis, safety first analysis, beta ratios, and expected value analysis have all been advocated. A useful reference on risk analysis is Anderson and Dillon [1992].

- Is it politically possible?
- Is it in line with national, regional and local developmental priorities?
- Is it socially acceptable in the sense of contributing to the cultural and social norms of society?
- Is it economically beneficial to the society at large?
- Is the institutional capacity potentially available to effectively implement the policy?

The final decision on the appropriateness of the 'model' and its components, will be made by the intended users according to whether they adopt or reject the recommendations. It is therefore worthwhile, as the second part of the preliminary appraisal, returning to the field and discussing the 'model' and its range of potential component enterprises and management practices with the involved parties, including the farmers. One or more village level participatory appraisal workshops should also be arranged for the team members to present their provisional recommendations, for assessment as to their suitability, to representatives of the farming community and other agencies operating within the target area. Farmers invited to participate in these workshops may be selected from among those interviewed during the informal survey. They should be innovative and responsive to new ideas and ones that can be relied on to give honest responses. Care should be taken to ensure that they are representative of the recommendation domain for which the 'model' is intended. Separate workshops may need to be arranged for different groups if there is more than one recommendation domain within the study area. It is important to appreciate that a non-committal or apparent positive response from farmers may be less valuable in the appraisal exercise than a strong negative response. The latter indicates that the team members have got something wrong and will need to rethink the overall 'model' or particular components of it.

Thus much of the appraisal at this stage involves 'back of the envelope calculations.' complemented and supplemented with information and opinions of other knowledgeable and involved people. As a result of this activity decisions are made as to what actions are passed onto the next phase, testing/implementation.

5.6 THE TESTING/IMPLEMENTATION PHASE

At this stage, representative farmers have 'hands-on' experience with testing/implementing the proposed solutions. This provides an opportunity ***ex post*** to obtain actual data (i.e., both quantitative and qualitative) for evaluating the proposed solutions. According to Table 4, three steps (i.e., C1.1-C1.3) are proposed at this stage. By now the following activities should have been completed for each of the identified recommendation domains:

- Present characteristics and performance of the 'typical' farming system of each recommendation domain to have been documented.
- Problems and constraints to have been identified and ranked in order of importance. Causes and effects also to have been determined and potential intervention points identified.
- Several alternative potential improvements to have been identified and technical and policy activities specified after appropriate appraisal.

5.6.1 Determine the Research Needs

The first activity in the testing/implementation phase is to determine the research needs of each proposed improvement.. This requires the team to review their current knowledge on the components of the 'model' farming systems. The purpose is to determine whether the improved enterprises, field level technologies and farm management practices that make up the 'model' are sufficiently tried and tested to be directly disseminated to farmers.

Alternatively is there a need to conduct research on some or all of them to test that they are suitable for dissemination? The more that is known about the technology, and the more that it has been subjected to practical farm level testing, the less is the need for further research. However, if little is known, initial research may need to be done at the RMRI level, probably on an experiment station, and then only subject to RMFI and FMFI level testing involving farmers once the cause/effect relationships are well understand. Because of the time required to derive research results, particularly if one has to start at the RMRI level, it would probably be desirable to find some other interim solution that can be implemented pending the availability of a better solution later as a result of research.

5.6.2 Research and Extension Planning

The next step is to develop an overall plan of action detailing research and extension programmes, and the linkage and interaction that should occur between the involved parties (e.g., farmers, researchers, extension staff, etc.). On the basis of this plan, it should be possible to refine estimates of the time frame and resources required to obtain results desired by the authorities. If necessary, the estimates can be submitted to the requisite authorities to enlist their support and possibly modify their expectations.

5.6.3 Implementation of Research and Extension Activities

At this point trials need to be implemented, preferably with heavy farmer involvement (i.e., in RMFI or FMFI format). Analysis and evaluation of the results can be undertaken using the same methods described earlier, except that now, actual rather than conjectural data, can be used. One deficiency of the present system of testing, which was discussed earlier (Section 4.4.2), is the difficulty of obtaining definitive information on long-term benefits and possible costs that relate to sustainability.[76] Participatory workshops/field discussions involving farmers are another very important way of evaluating results of trials, particularly involving those with which they have had a close association.

Once there is sufficient confidence that the proposed improvements are capable of addressing the production and sustainability problems of a specific recommendation domain, the next activity in the new approach to soil conservation is to disseminate the recommendations to farmers. It should be noted that there will be many occasions where the team will be able to identify management practices and technologies about which sufficient is already known, to satisfy fully the appraisal criteria outlined earlier, thus enabling them to be immediately recommended to farmers with a sufficient degree of confidence on their acceptability. When this is the case, no research and testing will be needed, and the team will be able to proceed directly to arranging for the dissemination of the proposed improvements.

Dissemination involves the team working with the local extension staff to develop the means for transferring to specific groups of farm households (i.e., recommendation domains), the knowledge required to enable them to evaluate the recommended components of a 'model' improved farming system, according to their household needs and objectives. The aim is to find ways to influence their decision making process so that they will perceive the improvements as beneficial and therefore worth adopting.

Providing implementation of the 'model' does not require the injection of resources not currently available locally (e.g., a credit programme, purchased

---

[76] The same problem applies, to an even greater extent, with respect to the implementation of physical structures designed to correct erosion problems. Benefit/cost analysis techniques, in principle, are appropriate analytical tools, but problems of predicting the future can raise some doubt about their validity.

inputs, additional extension personnel), it should be possible for the team members and the local extension staff to develop a flexible extension package that can be incorporated into the local ongoing farmer training extension programme. The more farmers can see and participate in implementing demonstrations of the suggested improvements, the better since, they, like all of us, are more likely to be convinced through 'doing' rather than simply observing.

Finally, if changes are required in the policy/support system to ensure adoption, then it may be necessary to postpone dissemination until such changes have been put in place. As we mentioned earlier, lack of congruency between technologies and policy/support systems, and between policies designed to stimulate production, and those designed to encourage sustainability, have been major reasons for the lack of success of many agricultural development initiatives.

5.7 DISSEMINATION IMPACT PHASE

According to Table 4, two steps (i.e., D1.1 and D1.2) are suggested in this phase. The first is concerned with monitoring progress on testing and disseminating the recommendations for improving the productivity and sustainability of the farming systems, and evaluating the reasons for their adoption, modification or rejection by the farm households of the study area. The second uses the results of the first as the basis for further analysis and development work to refine and improve the recommendations.

5.7.1 Monitoring and Evaluation

Monitoring and evaluation are interrelated processes concerned with the collection and analysis of information of on-going and completed activities. They are an essential part of the new approach to soil conservation since by documenting and analysing the results of the development work, they enable an assessment to be made as to whether the recommendations are successful in tackling the farm households' production and sustainability problems. If they are not, or not as successful as anticipated, then the monitoring and evaluation exercise should help identify why.

Monitoring involves the systematic collection of data in an organised manner to provide information on the implementation of the research and dissemination activities (i.e., implementation monitoring) and as the basis for ongoing and *ex post* evaluation (i.e., impact monitoring). Implementation monitoring provides management information to the team, and its supervisors, on the utilisation of research and extension inputs, on the timing of activities, and on the timely generation of outputs. Impact monitoring is more directed to effects, positive or negative, on the participating farm households and their objectives. Impact monitoring is concerned with collecting data for the purpose of enabling the team to detect and follow change within the farm household systems of the different recommendation domains.

Evaluation is undertaken using the data obtained from the implementation and impact monitoring exercises. These data are compared with the work plans and expected results (i.e., targets) of the research trials, and the dissemination activities, to see if they are being achieved. An evaluation into how successfully the planned research and extension activities were implemented may be positive, while an evaluation of the impact on farm households may be negative. In other words, the evaluation may reveal that although the team organised their work programme in an optimal manner, the inputs and proposed innovations did not result, as expected, in positive effects on the farm household production and sustainability problems.

In addition to this somewhat conventional approach in which the effect on the beneficiaries of a project are monitored and evaluated by 'outsiders' there is merit in trying to undertake monitoring and evaluation as a participatory exercise through including the beneficiaries themselves in the exercise. For example, participatory monitoring could involve groups of farm households, perhaps selected on a recommendation domain basis, collecting, processing and sharing data on their farming activities and the condition of their land resources, amongst themselves and with the team. Participatory evaluation would be the systematic analysis by the anticipated beneficiaries, with the assistance of the team members, of the monitored information, with a view to:

- Enabling them to adjust or refine particular field interventions.
- Identifying new farm household and community level problems and constraints requiring investigation.
- Recommending new research priorities in the light of their experiences in using the currently recommended technologies.
- Recommending changes and improvements to policies and institutional support services.

Participatory evaluation is expected to take the form of discussion sessions during participatory village level workshops while consolidating the farm level monitoring data. It can also take place during field visits while discussing observed changes in farming practices and the nature and severity of any visible land degradation. Evaluation done this way stimulates critical self-awareness amongst the farm households of the impact of their farming practices and can provide the motivation to make improvements.

Team members should use their participation in the village level workshops and field visits to promote a constant two way flow of information between themselves and the farmers. They should not use these solely as a way of extracting data, but should feed back their own findings so as to contribute to the village level analysis. However, this should be done without imposing their own conclusions. There are no hard and fast rules on how to conduct participatory monitoring and evaluation as it will depend on the confidence and experience of the team members, the rapport and trust they have been able to establish with the farmers, and the social and political structure within the community. However, by this time, the team will have had considerable exposure to the intended beneficiaries and therefore should have an idea on what methods would be possible.[77]

5.7.2 *Revision of Research and Extension Activities and Policy/Support Systems*

Although adoption or acceptability indices [Hildebrand and Poey, 1985] can be useful in assessing the degree of adoption, we have just indicated there are more informal and participatory methods that can elicit information useful in determining whether there is a need for feeding back requests for adjustments in research and extension activities and changes in policy/support systems that will improve levels of productivity and sustainability in a manner that is equitable. Since we are proposing more informal methods of obtaining this information, it would be highly desirable for representatives of research, extension and planning to participate directly in some of the village level workshops and field visits to hear directly from farmers about the problems. If this is not possible, then the only alternative, if their future cooperation is to be enlisted, is for more conventional approaches to be used involving collection of hard, quantitative data. This is obviously much more expensive and time consuming.

---

77 FAO [1988B] provides details on the role of monitoring and evaluation in small-scale farmer development, the selection and use of indicators and basic participatory monitoring and evaluation procedures.

6. IMPLEMENTING THE NEW SOIL CONSERVATION APPROACH

6.1 THE INSTITUTIONAL SETTING FOR IMPLEMENTATION

Some major concerns obviously exist about effectively implementing the new soil conservation approach. The interdisciplinary and holistic requirements of the approach demand skills not conventionally found in soil conservation specialists. Indeed, it would be unreasonable ever to expect all the skills required to be present in a single individual. Therefore, effective implementation of the approach will require soil conservation specialists to build up linkages and effective working relationships with representatives of other disciplines and in other institutions. In effect what is required, is "for the activities of the soil conservation specialists to be integrated into routine agricultural development activities of the field level" [Douglas, 1991B, p. 2]. What will be possible will tend to be very locational specific and will depend greatly on the linkages that are established. For example:

- ***Preventative Thrust.*** The optimal situation is likely to occur where FSD teams already exist. In such cases, incorporating the preventative soil conservation thrust (i.e., land husbandry) into their work programmes, should be relatively easy. In addition, two of the major principles of the new soil conservation approach can be addressed, namely:

    - A holistic approach is possible because the expertise of the different disciplines on the team can be tapped.

    - Because the 'felt' needs of farming households are already being addressed, strategies designed to ensure sustainability can be 'piggy-backed' onto strategies that yield short-term benefits.

Where no FSD teams exist, using a holistic approach and implementing preventative soil conservation type strategies will be more difficult. However, an attempt can be made to ensure that all research efforts designed to increase short-run production include sustainability criteria within their evaluation process.

- ***Corrective Thrust***. Currently, many soil conservation activities are located institutionally as separate units within the extension service. Development is required of effective working linkages with individual subject matter specialists and experienced personnel in the field (e.g., extension staff, community development specialists, staff in non-governmental organizations) and at the centre (e.g., researchers -- both technical and social -- planners, etc.). Certainly, such an approach decreases the possibility of soil conservation personnel having the more traditional narrow focus of dealing with the soil only and improves the possibility that practical strategies to combat the causes will be developed and implemented. Of course, in the rare situations in which farmers articulate their major concerns in terms of soil degradation problems, soil conservation staff can more easily have a positive impact, even without the sustained support of staff in other disciplines, as long as they are sensitive to the needs of the farmers.

- ***Policy Thrust***. As has been repeatedly emphasized, policy/support systems can be critically important in enhancing or inhibiting strategies to prevent or cure soil degradation. There are no easy answers to ensuring that the interactive linkages required for this to occur are created. However, the following steps could help.

  - Influence the appropriate governmental institutions so that there is much closer coordination/cooperation between those units responsible for policies/programmes dealing with production and conservation.

- Undertake studies evaluating adoption of soil conservation strategies and channel the results to those responsible for designing policies. If possible, in order to make these studies more palatable, enlist policy maker support (i.e., moral and/or preferably direct) prior to undertaking them.

- Ensure that planning and soil conservation interests are represented in any institutional arrangements for approving recommendations prior to dissemination to the extension service.

In any case, future training programmes for soil conservation specialists should contain components that will help them to communicate with farmers and sensitize them to the potential roles of other disciplines. It would also be desirable, whenever feasible, to provide them with at least some minimal skills in areas that are usually contributed by other disciplines.

6.2 CONCLUDING COMMENTS

In this publication, a case has been made for using the principles of FSD in implementing the new approach to soil conservation. These principles can potentially make a major contribution. However, as experience develops with applying them, there will naturally be refinements in the approach and techniques used. Applying the principles of FSD to the new soil conservation approach involves challenges. For example, two obvious problem areas are approaches to dealing with off-site costs of soil erosion and estimating the long-term benefits of soil conservation strategies within a reasonable time period. Nevertheless, a good start can be made. Soil conservation and agricultural production no longer should be regarded as separate activities. Good land husbandry is the key to effective soil conservation. Therefore, acceptance of the new soil conservation approach will depend on the greater integration of soil conservation staff with farmers, researchers, extension workers, and planners.

REFERENCES

Amir, P., and H. Knipscheer, 1989. Conducting On-Farm Animal Research: Procedures and Economic Analysis. Morrilton and Ottawa: Winrock International and International Development Research Centre.

Anderson, J., and J. Dillon, 1992. "Risk Analysis in Dryland Farming Systems." Farm Systems Management Series No. 2. Rome: Food and Agriculture Organisation.

Anderson, J., and J. Thampapillai, 1990. "Soil Conservation in Developing Countries: Project and Policy Intervention". Policy and Research Series Number 8. Washington: Policy, Research and External Affairs, The World Bank.

Andrew, C.O., and P.E. Hildebrand, 1982. Planning and Conducting Applied Agricultural Research. Boulder: Westview Press.

Bartlett, P.F., 1978. What Shall We Grow? A Critical Survey of the Literature on Farmers' Decision Making. Washington: Civic Participation Division, USAID.

Biggs, S.D., 1980. "Informal R and D." Ceres, 13(4): 76.

Binswanger, H.P., and V.W. Ruttan, 1978. Induced Innovation: Technology, Institutions and Development. Baltimore: John Hopkins.

Caldwell, J., and L. Walecka (Eds.), 1987. "Design Techniques for On-Farm Experimentation". FSR/E Training Units: Volume II. Second Edition. Gainesville: University of Florida, 1987.

Caldwell, J., D. Taylor, and L. Walecka (Eds.), 1987. "Analysis and Interpretation of On-Farm Experimentation". FSR/E Training Units: Volume III. Gainesville: University of Florida.

Caldwell, J., T. Frankenberger, T., D. Taylor, and L. Walecka (Eds.), 1987. "Diagnosis Design and Analysis in Farming Systems Research and Extension". FSR/E Training Units: Volume I, II and III Trainer's Manual. Gainesville: University of Florida.

CDCS, 1986. Soil and Water Conservation in Sub-Saharan Africa, Issues and Options. Amsterdam: Centre for Development Cooperation Services, Free University and International Fund for Agricultural Development.

Chambers, R., 1983. Rural Development: Putting the Last First. Harlow Longman.

Chambers R., 1985. "Shortcut Methods of Gathering Social Information for Rural Development Projects." In M. Cernea (Ed.) Putting People First. New York: Oxford University Press.

Chambers, R., 1992. "Rural Appraisal: Rapid, Relaxed and Participatory." Discussion Paper No. 311. Brighton: Institute of Development Studies, University of Sussex.

Chambers, R. and J. Jiggins, 1987. "Agricultural Research for Resource-Poor Farmers." Agricultural Administration and Extension, 27(2):109-128.

Chambers, R., R. Longhurst and A. Pacey (Eds.), 1981. "Seasonal Dimensions to Rural Poverty." London: Pinter.

Chambers, R., A. Pacey and L. A. Thrupp (Eds.), 1989. Farmer First: Farmer Innovation and Agricultural Research. London: Intermediate Technology Publications.

Cheatle R.J., 1993. "Next Steps Towards Better Land Husbandry." In Hudson N.W. and Cheatle R.J., (Eds.) "Working with Farmers for Better Land Husbandry." London: Intermediate Technology Publications.

CIMMYT, 1988. From Agronomic Data to Farmer Recommendations: an Economics Training Manual plus Workbook. Completely Revised Edition. Londres: Centro Internacional de Mejoramcento de Maiz g Trigo.

Conway, G.R., 1987. "The Properties of Agroecosystems." Agricultural Systems, 24:95-117.

Costin, A., and H. Coombs, 1981. "Farm Planning for Resource Conservation." Search, 12(12):429-430.

Davies, S., M. Leach and R. David, 1991. "Food Security and the Environment: Conflict of Complementarity?" Institute of Development Studies DP285. Brighton: IDS, University of Sussex.

Dillon, J. and J.B. Hardaker, 1980. "Farm Management Research for Small Farmer Development." FAO Agricultural Services Bulletin No. 41. Rome: Food and Agricultural Organisation of the United Nations.

Douglas, M., 1988. Integrating Conservation into the Farming System: Land Use Planning for Small-Scale Farmers: Concepts and Procedures. London: Food Production and Rural Development Division, Commonwealth Secretariat.

Douglas, M., 1990. "Integrating Conservation into the Farming System: Land Use Planning for Small-Scale Farmers." Soil Conservation Notes, 22: 3-11.

Douglas, M., 1991A. "Making Conservation Farmer Friendly." Paper given at WASWC/IIED Workshop on Soil and Water Management for Sustainable Agriculture, Taita Hills, Kenya and Arusha, Tanzania, 2nd-11th June, 1991.

Douglas, M., 1991B. "The Development of Conservation Farming Systems: Some Policy and Institutional Considerations." Keynote paper given at International Workshop on Conservation Policies for Sustainable Hillslope Farming, Solo, Indonesia, 11th-15th March.

Douglas, M., 1992. "A Framework for the Development of Conservation Effective Farming Systems." Produced for The Farm Management and Production Economics Service, Agricultural Services Division. Rome: Agricultural Services Division, Food and Agriculture Organization of the United Nations.

Douglas, M., 1993 Sustainable Use of Agricultural Soils. Discussion paper prepared for the Swiss Development Cooperation on behalf of the Group for Development and Environment, Institute for Geography, University of Bern, Switzerland.

Dumsday, R., 1971. "Evaluation of Soil Conservation by Systems Analysis." In Dent, J., and J. Anderson (Eds.), 1971. Systems Analysis in Agricultural Management. Sydney: Wiley.

Ellis, F., 1988. Peasant Economics. Cambridge: Cambridge University Press.

FAO, 1976. "A Framework for Land Evaluation." FAO Soils Bulletin No. 32. Rome: Food and Agricultural Organization.

FAO, 1983. "Guidelines: Land Evaluation for Rainfed Agriculture." FAO Soils Bulletin No. 52. Rome: Food and Agricultural Organization.

FAO, 1984. "Land Evaluation for Forestry." FAO Forestry Paper No. 48. Rome: Food and Agricultural Organization.

FAO, 1985. "Guidelines: Land Evaluation for Irrigated Agriculture." FAO Soils Bulletin No. 55. Rome: Food and Agricultural Organization.

FAO, 1986. "The Technology Application Gap: Overcoming Constraints to Small-Farm Development." Research and Technology Paper 1. Rome: Food and Agricultural Organization.

FAO, 1988A. "Guidelines: Land Evaluation for Extensive Grazing." FAO Soils Bulletin No. 58. Rome: Food and Agricultural Organization.

FAO, 1988B. "Participatory Monitoring and Evaluation, Handbook for Training Field Workers." RAPA Publication 1988/2. Bangkok: FAO Regional Office for Asia and the Pacific.

FAO, 1989. Farming Systems Development: Concepts, Methods, Applications. Rome: Agricultural Services Division, Food and Agriculture Organization of the United Nations.

FAO, 1990A. Farming Systems Development: Guidelines for the Conduct of a Training Course In Farming Systems Development. Rome: Agricultural Services Division, Food and Agriculture Organization of the United Nations.

FAO, 1990B. "The Conservation and Rehabilitation of African Lands: an International Scheme." ARC/90/4. Rome: Food and Agricultural Organization.

FAO, 1992. Protect and Produce -- Putting the Pieces Together. Rome: Food and Agricultural Organization.

Farrington, J., and A. Martin, 1988. "Farmer Participation in Agricultural Research: a Review of Concepts and Practices." Agricultural Administration Occasional Paper No. 9. London: ODI.

Feldstein, H.S., and S.V. Poats (Eds.), 1989. Working Together. West Hartford: Kumarian Press.

Francis, C.A. and P.E. Hildebrand, 1989. "Farming Systems Research/Extension and the Concepts of Sustainability." Paper presented at The Ninth Annual International Farming Systems Symposium, Fayetteville, University of Arkansas, 8th - 11th October.

Frankenberger, T.R., 1991. Personal Communication.

Frankenberger, T.R., and D.M. Goldstein, 1990. "Food Security, Coping Strategies, and Environmental Degradation. Arid Lands Newsletter, 30:21-27.

Frankenberger, T.R., and D.M. Goldstein, 1992. "The Long and Short of it: Household Food Security, Coping Strategies, and Environmental Degradation in Africa. In Food for the Hungry Int (Eds.). Growing our Future: Food Security and the Environment. West Hartford: Kumarian Press. (Forthcoming).

Frankenberger, T.R., and C.F. Hutchinson, 1991. "Sustainable Resource Management Based on a Decentralized Food Security Monitoring System." Discussion Paper. Tucson: Office of Arid Lands Studies, University of Arizona.

Frankenberger, T., and L. Walecka (Eds.), 1987. "Diagnosis in Farming Systems Research and Extension". FSR/E Training Units: Volume I. Second Edition. Gainesville: University of Florida.

Franzel, S.C., 1984. "Modelling Farmers' Decisions in a Farming Systems Research Exercise: the Adoption of an Improved Maize Variety in Kirinyaga District, Kenya." Human Organization, 43(3):199-207.

Fujisaka, S., 1989. "A Method for Farmer-Participatory Research and Technology Transfer: Upland Soil Conservation in the Philippines." Experimental Agriculture, 25: 423-433.

Gallacher, R.N., 1987. The Search for Low-Input Soil and Water Conservation Techniques: Concepts and Trends. Rome: Land and Water Development Division, Food and Agricultural Organization.

Harrington, L.W., 1991. Measuring Sustainability: Issues and Alternatives. Paper given at the 11th Annual AFSRE Symposium, East Lansing, Michigan State University, 5th-10th October.

Hart, R., and M. Sands, 1991. "Sustainable Land-Use Systems Research and Development." Association For Farming Systems Research and Extension Newsletter, 2(1):1-6.

Haverkort B., van der Kamp J., and A. Waters-Bayer (Eds.), 1991. Joining Farmers' Experiments -- Experiences in Participatory Technology Development. London: Intermediate Technology Publications.

Hayami, Y., and V. Ruttan, 1971. Agricultural Development: an International Perspective. Baltimore: John Hopkins University Press.

Heinrich, G., S. Masikara, S. Magalela and G. Moremedi, 1990. "Testing an Accelerated Extension Approach in North East District: the Case of Extension-Oriented Farmer Groups." ATIP Working Paper 25. Gaborone: Department of Agricultural Research.

Heinrich, G., J. Siebert, E. Modiakgotla, D. Norman and J. Ware-Snyder (Eds.), 1990. "Technical Summary of ATIP's Activities, 1982-90: Research Results." ATIP RP 5. Gaborone: Department of Agricultural Research.

Hildebrand, P.E., 1989. "Agronomy's Role in Sustainable Agriculture: Integrated Farming Systems." Paper presented at The American Society of Agronomy Meetings, Las Vegas, Nevada; 16th October.

Hildebrand, P., and F. Poey, 1985. On-Farm Agronomic Trials in Farming Systems Research and Extension. Boulder: Lynne Rienner.

Hilsum L., 1987. The Terraces of Wollo. In Towards Sustainable Development. Panos Institute London.

Hudson, N.W., 1987. "Soil and Water Conservation in Semi-Arid Areas." FAO Soils Bulletin 57. Rome: Food and Agricultural Organization.

Hudson, N.W., 1988. "Tilting at Windmills or Fighting Real Battles." In Moldenhauer, W.C. and N.W. Hudson (Eds.) Conservation Farming on Steep Lands. Ankeny, Iowa: Soil and Water Conservation Society, World Association of Soil and Water Conservation. pp. 3-8.

Hudson N.W., 1991. "A Study of the Reasons for Success or Failure of Soil Conservation Projects." FAO Soils Bulletin No.64. Rome: FAO.

Hudson N.W., 1992. Land Husbandry. London: Batsford.

Hudson N.W. and R.J Cheatle (Eds.), 1993. Working with Farmers for Better Land Husbandry. London: Intermediate Technology Publications.

ICRAF, 1983. "Resources for Agroforestry Diagnosis and Design. WP No. 7. Nairobi: ICRAF.

ICRISAT (Eds.), 1986. Farming Systems Principles for Improved Food Production and the Control of Soil Degradation in the Semi-Arid, and Humid Tropics. Hyderabad: United Nations Environment Programme and International Crops Research Centre for the Semi-Arid Tropics.

IFAD, 1992. Soil and Water Conservation in Sub-Saharan Africa. Towards Sustainable Production by the Rural Poor. Rome: International Fund for Agricultural Development.

Jodha, N.S., 1990. "Sustainable Agriculture in Fragile Resource Zones: Technological Imperatives." Paper presented at The International Symposium on Natural Resource Management for a Sustainable Agriculture, New Delhi, 6th-10th October.

Kearl, B. (Ed.), 1976. Field Data Collection in the Social Sciences: Experiences in Africa and the Middle East. New York: Agricultural Development Council.

Khon Kaen University (Eds.), 1987. Proceeding of the 1985 International Conference on Rapid Rural Appraisal. Khon Khaen: Rural Systems Research and Farming Systems Research Projects, Khon Kaen University.

Magrath, W. and P. Arens, 1989. "The Costs of Soil Erosion on Java: A Natural Resource Accounting Approach." Environment Department Working Paper No. 3. Washington: The World Bank.

Mascarenhas, J., P. Shah, S. Joseph, R. Jayakaran, J. Devavaram, V. Ramachandran, A. Fernandez, R. Chambers, and J. Pretty (Eds.), 1991. "Participatory Rural Appraisal. Proceedings of the February 1991 Bangalore PRA Trainers Workshop." RRA Notes No. 13. London and Bangalore: IIED and MYRADA.

McCorkle, C.M., 1989. "Toward a Knowledge of Local Knowledge and Its Importance for Agricultural RD and E." Agriculture and Human Values, 6(3): 4-12.

Moldenhauer, W.C. and N.W. Hudson (Eds.), 1988. Conservation Farming on Steep Lands. Ankeny, Iowa: Soil and Water Conservation Society, World Association of Soil and Water Conservation.

Molnar, A., 1989. "A Review of Rapid Appraisal Tools for use in Natural Resource Management Planning and Project Design and Execution." A review paper for FAO Forestry Department, Rome: Food and Agricultural Organization.

Muller, E.U., and S.J. Scherr, 1990. "Planning Technical Interventions in Agroforestry Projects." Agroforestry Systems, 12: 23-44.

Napier, T., 1987. "Farmers and Soil Erosion: A Question of Motivation." Forum, 2(2): 85-94.

Napier, T., 1989. "Implementation of Soil Conservation Practices: Past Efforts and Future Prospects." Topics in Applied Resource Management, 1: 9-34.

Napier, T. and D. Forster, 1982. "Farmer Attitudes and Behaviour Associated with Soil Erosion Control." In Halcrow, H.G., E.O. Heady and M.L. Cotner (Eds.), Soil Conservation Policies and Incentives. Ankeny, Iowa: Soil Conservation Society of America.

Norman, D.W., 1982. "Socioeconomic Considerations in Sorghum Farming Systems." In ICRISAT (Ed.), Sorghum in the Eighties. Andhra Pradesh: International Crops Research Centre for the Semi-Arid Tropics. pp. 633-646.

Norman, D., 1989. "Accountability: a Dilemma in Farming Systems Research." Culture and Agriculture, 38 (Spring/Summer): 2-8.

Norman, D., 1991. "Soil Conservation: Using Farming Systems Development as an Aid." Produced for The Farm Management and Production Economics Service, Agricultural Services Division. Rome: Agricultural Services Division, Food and Agriculture Organization of the United Nations.

Norman, D., and M. Collinson, 1986. "Farming Systems Research in Theory and Practice." In J. Remenyi (Ed.), Agricultural Systems Research for Developing Countries, ACIAR Proceedings No. 11. Canberra: Australian Council for International Agricultural Research. pp. 16-30.

Norman, D., and D. Pratt, 1990. "Harnessing Farming Systems Concepts and Methods for Community and Area Development: Institutionalizing Farming Systems Development." Consultants Report. Rome: AGSP, Food and Agricultural Organization.

Norman, D., E. Simmons, and H. Hays, 1982. Farming Systems in the Nigerian Savanna: Research and Strategies for Development. Boulder: Westview Press.

Pimentel, D., E. Terhune, R. Dyson-Hudson, S. Rochereau, R. Samis, E. Smith, D. Denman, D. Reifschneider and M. Shepard, 1976. "Land Degradation: Effects on Food and Energy Resources." Science, 194: 149-155.

Pretty, J., 1989. "A Game-Centred Approach to Development: Games, Groups and Rapid Rural Appraisal." International Institute for Environment and Development (IIED). Draft for Rome: Food and Agricultural Organization.

Prior J.C., 1992. "Evaluating the Experience of Third World Soil and Water Conservation Projects -- Nine Common Shortcomings." Proceedings of the 7th International Soil Conservation Conference. Sydney: Soil Conservation Society.

Raintree, J.B., (Ed.), 1987. D&D User's Manual: An Introduction to Agroforestry Diagnosis and Design. Nairobi: International Council for Research in Agroforestry.

Richards, P., 1985. Indigenous Agricultural Revolution. London: Hutchinson.

Rocheleau, D.E., 1985. "Land Use Planning with Rural Farm Households and Communities: Participatory Agroforestry Research." Working Paper No. 36. Nairobi: International Council for Research in Agroforestry.

Rocheleau, D.E., 1990. "Participatory Research in Agroforestry Exploitation, Eclipse or Empowerment of Rural People's Science and Practice?" Invited paper at ICRAF International Workshop on Methods for Participatory On-Farm Agroforestry Research, 19th-23rd February, Nairobi: International Council for Research in Agroforestry.

Rhoades, R., and R.H. Booth, 1982. "Farmer-back-to-Farmer: A Model for Generating Acceptable Agricultural Technology." Agricultural Administration, 11: 127-137.

Ruttan, V., 1988. "Sustainability is not Enough." American Journal of Alternative Agriculture. 3: 128-130.

Sanders, D. W., 1988. "Soil and Water Conservation on Steep Lands: a Summary of Workshop Discussions." In Moldenhauer, W. C. and N. W. Hudson (Eds.) Conservation Farming on Steep Lands. Ankeny,

Iowa: Soil and Water Conservation Society, World Association of Soil and Water Conservation. pp. 275-282.

Shaner, W.W., P.F. Philipp, and W.R. Schmehl, 1982. Farming Systems Research and Development: Guidelines for Developing Countries. Boulder: Westview Press.

Shaxson, T.F., 1988. "Conserving Soil by Stealth." In Moldenhauer, W. C. and N. W. Hudson (Eds.) Conservation Farming on Steep Lands. Ankeny, Iowa: Soil and Water Conservation Society, World Association of Soil and Water Conservation. pp. 9-17.

Shaxson T.F. 1989. "Conservation Farming Through People's Participation". Final Report of Project TCP/LES/6755 Development of a Watershed Management Strategy. Maseru: Government of Lesotho and the Food and Agricultural Organisation.

Shaxson, T., N. Hudson, D. Sanders, E. Roose, and W. Moldenhauer, 1989. Land Husbandry: a Framework for Soil and Water Conservation. Ankeny, Iowa: Soil and Water Conservation Society, World Association of Soil and Water Conservation.

Sheng T.C. 1989. Soil conservation for Small Farmers in the Humid Tropics. FAO Soils Bulletin No. 60, Rome.

Singh, I., B. Singh, and H. Bal, 1987. "Indiscriminate Fertilizer Use vis-a-vis Ground Water Pollution in Central Punjab." Indian Journal of Agricultural Economics 42(3): 404-409.

Stocking, M., 1986. "The Cost of Soil Erosion in Zimbabwe in Terms of the Loss of Three Major Nutrients." Consultants' Working Paper No 3. Rome: AGLS, Food and Agricultural Organization.

Stroup, W., P. Hildebrand and C. Francis, 1991. "Farmer Participation for More Effective Research in Sustainable Agriculture." Staff Paper SP91-32. Gainesville: Food and Resource Economics Department, University of Florida.

Tripp, R., and J. Woolley, 1989. The Planning Stage of On-Farm Research: Identifying Factors for Experimentation. Londres and Cali: Centro de Internacional de Mejoramcento de Maiz g Trigo and Centro Internacional de Agricultura Tropical.

van Vuuren, W., 1986. "Soil Erosion: the Case for Market Intervention." Canadian Journal of Agricultural Economics, 33(Annual Meeting and Workshop Proceedings): 41-62.

Walters, D., D. Mortensen, C. Francis, R. Elmore and J. King, 1990. "Specificity: The Context of Research for Sustainability." The Journal of Soil and Water Conservation, 45(1): 55-57.

Watts, M., 1988. "Coping with the Market: Uncertainty and Food Security among Hausa Peasants." In De Garine, I., and G.A. Harrison (Eds.), Coping with Uncertainty in Food Supply. Oxford: Clarendon Press. pp. 260-290.

Williams, J.R., 1990. "Social Traps and Incentives: Implications for Low-Input Sustainable Agriculture." The Journal of Soil and Water Conservation, 45(1): 28-30.

World Bank, 1986. "Land and Water Resources Management Seminar." Washington DC: Economic Development Institute, International Bank for Reconstruction and Development.

Worman, F., D. Norman and J. Ware-Snyder (Eds.), 1990. "Farming Systems Research Handbook for Botswana." ATIP RP 3. Gaborone: Department of Agricultural Research.

Zandstra, H., E. Price, J. Litsinger, and R. Morris, 1981. A Methodology for On-Farm Cropping Systems Research. Los Banos: International Rice Research Institute.

Zandstra, H., K. Swanberg, C. Zulberti and B. Nestel, 1979. "Caqueza: Living Rural Development." International Development Research Centre IDRC - 107e. Ottawa: IDRC.